Franz Kurfeß

# Parallelism in Logic

# Artificial Intelligence

# Künstliche Intelligenz

edited by Wolfgang Bibel and Walther von Hahn

Artificial Intelligence aims for an understanding and the technical realization of intelligent behaviour.
The books of this series are meant to cover topics from the areas of knowledge processing, knowledge representation, expert systems, communication of knowledge (language, images, speach, etc.), AI machinery as well as languages, models of biological systems, and cognitive modelling.

*In English:*

**Automated Theorem Proving**
by Wolfgang Bibel

**Parallelism in Logic**
by Franz Kurfeß

**Relative Complexities of First Order Calculi**
by Elmar Eder

*In German:*

**Die Wissensrepräsentationssprache OPS 5**
by Reinhard Krickhahn and Bernd Radig

**Prolog**
by Ralf Cordes, Rudolf Kurse, Horst Langendörfer and Heinrich Rust

**LISP**
by Rüdiger Esser and Elisabeth Feldmar

**Logische Grundlagen der Künstlichen Intelligenz**
by Michael R. Genesereth and Nils J. Nilsson

**Wissenbasierte Echtzeitplanung**
by Jürgen Dorn

**Modulare Regelprogrammierung**
by Siegfried Bocionek

**Automatisierung von Terminierungsbeweisen**
by Christoph Walther

**Logische und Funktionale Programmierung**
by Ulrich Furbach

Franz Kurfeß

# Parallelism in Logic

Its Potential for Performance
and Program Development

Die Deutsche Bibliothek – CIP-Einheitsaufnahme

**Kurfess, Franz:**
Parallelism in logic: its potential for performance
and program development / Franz Kurfess. – Braunschweig:
Vieweg, 1991
   (Artificial intelligence)

AMS Subject Classification: 03 B 70, 68 N 05, 68 M 20, 68 Q 10, 68 T 25

Verlag Vieweg · P. O. Box 58 29 · D-6200 Wiesbaden · FR Germany

Vieweg is a subsidiary company of the Bertelsmann Publishing International.



Cover design: L. Markgraf, Wiesbaden

ISBN-13: 978-3-528-05163-1    e-ISBN-13: 978-3-322-84922-9
DOI: 10.1007/ 978-3-322-84922-9

# Preface

The work described in this book is the outcome of the author's contributions to the design of a parallel inference machine in the Artificial Intelligence / Intellektik Group at the Computer Science Department of the Technical University in Munich. It is the revised version of a doctoral dissertation, submitted to the Technical University under the same title in 1990. The research was carried out in the framework of the European ESPRIT programme, project 415: "Parallel Architectures and Languages for Advanced Information Processing – A VLSI-directed Approach", with Philips Research Laboratories, Eindhoven, The Netherlands, as main contractor, and Nixdorf Computer AG, Paderborn, Germany, as subcontractor for subproject F.

Out of the many people who helped me to accomplish this work, I owe the most to Prof. Dr. W. Bibel, who founded the Intellektik Group in Munich, initiated its participation in ESPRIT 415, and hired me for this project although my exposure to logic was rather limited at that time. My thesis advisor, Prof. E. Jessen, deserves a lot of gratitude for his superb supervision and constant encouragement, especially after Prof. Bibel left Munich; without his help, it would have taken me even longer to complete this work. Many more people contributed to the research described here; an overview can be found in the back of the book (p. 269).

My final thanks go to Catharine O'Shaughnessy for her patience and support, especially during the final stage.


Berkeley, March 1991 Franz Kurfeß

## *Abstract*


This book discusses the potential of parallelism in logic programming and its exploitation on parallel architectures. Whereas the state of the art is almost exclusively restricted to an exploitation of AND- or OR-parallelism, a variety of *categories of parallelism* is discussed with respect to different levels of a logical formula (formula, clause, literal, term) and different ways to evaluate it. This discussion is based on formal specifications of the categories using the language UNITY (Chandy and Misra, 1988).

As an outcome of these investigations, a proposal is made to capture both the specification of parallel problems with logic, as well as the exploitation of parallelism in the evaluation of a logic formula or program: *Modularity* allows structuring of logic programs in an appropriate way, and *meta-evaluation* can be used to control the evaluation process on a parallel system. This combination yields a consistent programming framework with a wide scope ranging from program specification and verification to implementation and execution on parallel architectures.

Finally, the suitability of a specific evaluation mechanism for parallel architectures is investigated. The *spanning setter* concept is derived from the connection method (Bibel, 1987) and relies on a statical analysis of the formula to achieve largely independent computation units suited for execution on parallel computer systems.

The potential of parallelism in logic reaches far beyond the exploitation of AND- and OR-parallelism usually found in attempts to parallelize PROLOG. Moderate language enhancements and an appropriate evaluation mechanism offer a programming framework with strong expressiveness and efficient execution, without sacrificing the cleanliness of the underlying theoretical foundations.

# Contents

# List of Figures

# Chapter 1

# Introduction

The purpose of this introductory chapter is to provide some information on the background of this book. It sketches important aspects of logic, parallelism, and the combination of these two areas. It is neither an introduction to logic, nor to parallel systems, and the context of the work is mainly outlined by brief remarks and references to established textbooks and publications. Furthermore, this chapter contains a section illuminating the relevance of the work presented, with respect to important aspects of the overall area of computer science. The final section sketches the organizational structure of the book.

## 1.1  Logic

Drawing conclusions from basic facts is a very essential human skill, both in everyday life and professional arguing. Whereas *reasoning* comprises the relatively informal act of drawing conclusions, (mathematical) *logic* offers a formal mechanism for the derivation of conclusions according to a validated methodology, its classical instance being first order predicate logic. The goal of *reasoning* – in contrast to formal, mathematical logic – is to achieve conclusions without a highly sophisticated, rigid logic formalism. It is based on a more or less generally accepted methodology of connecting pieces of knowledge at hand, and deriving new items from them. Typical illustrations for this methodology can be found in many detective stories, where the agent makes use of scattered hints and evidences – which in principle are accessible to everybody – to "magically" derive his (or her) solution of the case under investigation.

The underlying rules of human reasoning, however, are only more or less generally agreed upon, and in the course of history, many a man may have suffered severely from such minor differences in the ways of drawing conclusions. *Mathematical* or *formal logic*, as the result of maybe two millennia's work of logicians, offers such a framework based on exactly defined, rigid rules for drawing conclusions. The power of logic is based on three important features: First, it provides a language for the accurate *representation of knowledge*, as far as this knowledge can be described by formal means. Second, a framework for *processing the represented knowledge* is given in the form of a calculus defining permitted rules for drawing conclusions. Third, a formal tool to mechanically derive *proofs* of the truth or equivalence of statements can be defined. The expressiveness of these formal systems is only limited by the descriptive power of formal systems (here well-defined formulae); problems, however, lie in the adequateness of the language (its coincidence with the human perception of a problem and the interpretation of a problem description), the efficiency of the evaluation mechanism, and the ability to deal with uncertainty, inconsistencies and incomplete knowledge usually present in real-world problems.

Since the advent of computing machinery these devices have been used to try to unload the burden of those stupid, mechanical steps in the proof process from the back of the plagued logician. The realization of systems for logic and reasoning has been pursued mainly in the field of Artificial Intelligence (AI), especially through systems manipulating symbols ("symbolic computation"); some agreeable results have been produced this way, and with PROLOG a programming language based on logic has become quite popular. But still automated theorem provers and problem solvers are either restricted to specialized fields, or suffer from poor performance.

A lot of effort in the formal logic field has been aimed at the creation of an adequate representation language allowing for a natural and comprehensible description of real-world knowledge, and the construction of a "smart" evaluation mechanism. Higher efficiency in the computation can be achieved by reducing the search space via domain knowledge (appropriate strategies), experience (learning from previous work), and by taking advantage of information on global and local features of the proof process. Reasoning systems, incorporating "sloppy" (untidy, common-place) features, strive for applicability to less formally characterized problems of knowledge representation and information processing by probabilistic evaluation methods, nonmonotonic calculi, flexible propositions or inference rules, and sloppy unification of terms.

Important milestones in the development of *logic* from its early beginnings to its present status can be found (among many others) in (Kneale and Kneale, 1984), in the work of (Aristoteles, 1976b; Frege, 1879; Whitehead and Russell, 1910), as well as in numerous textbooks, as for example (Smullyan, 1968; Hermes, 1976; Pospesel, 1976; Pospesel, 1984; Schöning, 1987); recent developments are reflected in a number of journals and conferences. The field of *automated theorem proving* is described in (Chang and Lee, 1973; Loveland, 1979; Boyer and Moore, 1979; Wos et al., 1984; Bibel, 1987; Bläsius and Bürckert, 1987); (Siekmann and Wrightson, 1983) contains classical papers on computational logic; major periodicals are *Journal of Automated Reasoning, Journal of Symbolic Computation* and the *Conference on Automated Deduction (CADE)* is the outstanding international colloquium in the field. The use of logic as a tool for programming computers, or *logic programming*, is mainly present in the language PROLOG ; sources of information on this language and resolution as underlying calculus are (Colmerauer et al., 1973; Robinson, 1979; Kowalski, 1979; Clocksin and Mellish, 1984; Bratko, 1986). Important periodicals are the *Journal of Logic Programming*, scientific meetings are *Conference on Logic Programming* and *Symposium of Logic Programming*. Many relevant articles can also be found in the *Artificial Intelligence* journal and in conferences like *International Joint Conference on Artificial Intelligence (IJCAI)*, *American Association for Artificial Intelligence (AAAI)*, and *European Conference on Artificial Intelligence (ECAI)*.

## 1.2   Parallelism

Although any computer system consists of a multiplicity of functional units on various levels, conventional computers as well as most programming languages are mainly viewed by the user and programmer as inherently *sequential*. The use of parallelism is initiated by a desire for higher performance in the first place; at a second glance, however, parallelism also offers a potential to describe inherently parallel problems or systems in an appropriate way, which may lead to better solutions through a deeper understanding of the problem to be solved. The first aspect is reflected by viewing a computer system as a *parallel execution vehicle*, executing more than one instruction or program simultaneously; the second provides a formal tool for the *specification of parallel problems*. Both aspects are unified by the notion of a *parallel program*: It can be used to describe parallel problems, and it may be executed on parallel systems. Further aspects of parallelism lie in the determination of different solutions for a problem, various approaches to find a solution, and a better exploitation of available resources (e.g. distributed computation in a network of workstations).

### Parallel Problem Specification

An adequate description of parallel problems must provide means to specify important aspects of the problem under investigation. The *structure* of the problem is implied by functional dependencies between sub-problems; these relations can be static, i.e. they remain unchanged throughout the whole processing of the task, or they may change dynamically. The structure can be described in an explicit specification, with graphs being widely used for static cases; for dynamically changing structures, the additional temporal dimension must be integrated, which can be obtained through a sequence of statical structures (graphs), or by instructions how to modify the initial structure. The alternative to an explicit specification is to reflect the structure of the problem implicitly in the structure of the program; this approach bears the danger of missing clarity for large or complex problems, especially by intermingling specification and implementation issues.

Furthermore, essential features regarding the *functionality* of (sub-)-problems must be described. This is basically the same as in the sequential case, differing only in the possible interactions between subproblems. In a sequential system, the processing of a problem must be mapped onto a single thread of execution, thus providing a rigid framework for interaction between subproblems. A parallel system features multiple threads of action, possibly leading to problems like faulty access to data or resources, or deadlock situations. Interaction between subproblems occurs in the form of *communication*, aiming at the exchange of data, or *synchronization*, which coordinates the processing of the different subproblems. Communication can be provided by shared data, where access to shared data must be possible for the subproblems involved in the exchange of information, and measures must be taken to guarantee the consistency of the data. An alternative to sharing data is the exchange of messages between subproblems; in this case, there must be a way of delivering messages from one subproblem to another. Synchronization can be achieved implicitly through the structure of the program (procedure calls, subroutines), access to shared data, or by explicit constructs like semaphores or monitors.

## Parallel Execution of Programs

Independent of a problem to be solved, the program describing how to process this problem can be executed in different ways. In a sequential execution, one path through the program is chosen, and interactions between subprograms are resolved by suspending one subprogram, treating the next, and resuming the former again. In addition, features like multiprocessing are used to provide quasi-parallel program execution. Parallel program evaluation requires a transformation of the program text in executable pieces of code (e.g. processes, objects), their allocation to processing elements, and the actual execution on the processing elements together with the interactions as specified in the program.

Throughout this book, there will be a differentiation between the evaluation of a logical formula and the execution of a logical program:

*evaluation* denotes the computation of a solution to a problem described by a logical *formula*; *execution* concentrates on carrying out the instructions specified by a logical *program*, and does not necessarily deliver the result of a computation (e.g. operating system)[1]. Since parallel evaluation and execution of programs tend to be quite complex actions, some intermediate stages are introduced. A *computational model* specifies the evaluation of programs according to a certain paradigm (e.g. procedural, object-oriented, functional, logic), whereas an *abstract machine* comprises the definition of instructions to be carried out during the execution of programs together with idealized resources on which execution is performed.

## Parallel Computer Architectures

Truly parallel execution of programs must be performed on computer system consisting of multiple processing elements. Important aspects of a parallel computer system are its structure, its operational behavior, and the functionality of its single processing elements. Essential for the *structure of a parallel system* are its organization (central / distributed memory, serial / parallel interconnections, bus / switching network / links), its topology (fully / partially interconnected; e.g. hypercube, torus, grid, ring, star), together with the identity of the processing elements it consists of (homogeneous / heterogeneous). The *operational behavior* is characterized by the mode of operation (synchronous / asynchronous), the exchange of information (shared data / message passing), and the synchronization mechanisms (general data structures / specialized data structures / hardware). The *functionality of the processing elements* is given by their intended usage (general / special purpose), their internal structure (CISC / RISC), and their size (small / large). A typical processing element comprises a computation unit, some local memory, and a communication unit.

Some books describing aspects of parallel programming are (Ben-Ari, 1984; De Bakker et al., 1986; Chandy and Misra, 1988); parallel computer systems and architectures are the subject of (Giloi, 1981; Hwang and Briggs, 1984; Treleaven and Vanneschi, 1987a; Kober, 1988; Gonauser and Mrva, 1989). Major journals in this area are *Par-*

---

[1] In some cases, however, the distinction between formula and program, or evaluation and execution, is not very essential, and the respective terms are used as synonyms.

*allel Computing*, *Distributed Computing*, and *Journal of Parallel and Distributed Computing*. In addition, recent information can be found in a number of journals and conferences on computer architecture, programming techniques and algorithm analysis.

## 1.3   Parallelism and Logic

Combining the descriptive power of logic with the potential of high performance parallel systems has become quite popular in the last few years. especially with the launch of the Japanese *Fifth Generation Computing* project pursued under the auspices of ICOT (Goto et al., 1988). This combination has mutual advantages for both areas involved: logic fertilizes parallel programming by its expressiveness, which allows the description of complex problems in a precise, but still quite natural way; parallel systems, on the other hand, provide a sufficiently powerful execution platform for the evaluation of logic formulae and programs. The use of logic as underlying programming paradigm for parallel computer systems also facilitates a clear separation between the *declarative* aspects of a problem to be specified, and the *imperative* aspects of an efficient execution of the resulting program.

**Problem Specification**

The description of complex problems and parallel systems with the means of logic must provide a basic set of fundamental mechanisms to capture the essential phenomena occurring in these systems. With particular attention to *parallel* systems, three categories can be identified:

- Representation; the *functional entities* of the problem or system to be described must be represented in an adequate way. Traditional representation mechanisms are data structures; the means logic provides for this purpose are constants, variables, and terms.

- Computation; the actions to be performed by or on the functional entities are mirrored in their *behavior*. This behavior is modeled with imperative means by manipulations of data structures based on (destructive) assignments to variables; in logic, predicates describe in a declarative way relations between the functional entities. This description by its very nature is statical, and must be combined with an evaluation mechanism to actually perform computation; an evaluation mechanism for logic program is derived from a logical calculus and should satisfy the criteria of correctness and completeness.

- Communication; especially in parallel systems the *exchange of information* between the single functional entities is very essential. Traditional ways to exchange information are the sharing of data, or the copying of data. In logic, two communicating items must come to a mutual agreement by unification.

In most parallel programming paradigms (including the ones parallelizing PROLOG), the description of a complex problem resides on a mixture of declarative (essential properties of an entity) and imperative (actions to be performed on or by an entity) features. Such a description cannot be achieved without a detailed knowledge of the underlying execution mechanism, whose operation typically is out of the user's control. In logic, metaprogramming concepts allow the adaptation of the execution mechanism to particular needs, resulting in a parallel programming paradigm with explicit execution control on the user's side while maintaining a clear separation of declarative and imperative aspects. Thus, the gap between the description of a parallel system and its implementation can be reduced substantially.

### Program Execution

The execution of programs written in a language based on logic is achieved by an implementation of a calculus; the calculus specifies the rules how to evaluate the relations described by the predicates of the program. Such an implementation usually is done either in an interpreted way, where the logic program is directly evaluated, or in a compiled way, where the logic program is translated into intermediate code which then is executed. This compiled evaluation process can be subdivided into three stages:

- transformations of the program to reduce the search space to be traversed;
- compilation into abstract machine instructions;
- execution of these instructions on a real machine.

In parallel logic systems, the compiled approach usually is favored since it allows for higher performance in the single processing elements and for better adaptation to the system configuration.

## Sources of Parallelism

The process of treating a problem with a computer system shows potential parallelism on three levels:

- Problem specification; on this level, the parallelism inherent in the problem to be described becomes apparent and should be reflected in the structure of its specification (e.g. through modules, objects, abstract data types). Different specifications of one and the same problem typically result in different degrees of parallelism.

- Formula evaluation; this level reveals parallelism inherent in the way how to evaluate a formula and depends to a certain degree on the underlying calculus. With the same formula as initial point, various calculi, or various strategies applied within one calculus, lead to different degrees of parallelism. If a compiled approach is taken, variations on the formula evaluation level yield different programs to be executed.

- Program execution; on this level, mainly three sources of parallelism can be identified. The first is the simultaneous execution of independent programs, the second the execution of multiple instructions of one and the same program, and the third the parallelization of a single instruction.

The potential of parallelism thus is governed by the problem definition and its formal specification, by the method used to evaluate the specification, and by the actual program executed on a particular system.


## Control of Parallelism

In order to efficiently compute a solution to a problem in a parallel way it is essential to identify the potential of parallelism to be exploited on the levels described above. It additionally involves the difficult task of adapting the characteristics of the computation process to the features of a particular (parallel) system on which the computation is to be executed. In accordance to the threefold distinction introduced above, measures and mechanisms for the control of parallelism are discussed for these levels.

- Problem specification; during the specification phase, important decisions about the structure of the problem description have to be made; the relations between the single components of the whole problem can be regular or irregular, they may remain statically the same, or may change dynamically. Of extreme importance on this level are the paradigm chosen for the specification (e.g. object-oriented, functional, logical), and knowledge about the evaluation method applied on the next stage is of advantage.

- Formula evaluation; on the evaluation level, the control of parallelism is correlated with (the variant of) the calculus applied; this level is essential for the origin of parallelism. Important aspects here are alternative computations during the evaluation process, the structure of the evaluation process (visualize the proof tree for a logical formula), the creation, deletion or modification of data items and computational tasks (e.g. processes), the grain size of the single computational tasks, communication mechanisms, and synchronization constructs. Decisions on this level have a strategic bearing on the execution level, and knowledge about the actual execution mechanism and its underlying hardware is highly favorable for efficient execution.

- Program execution; the main problem on the execution level is the balance between the parallelism evoked on the evaluation level and the parallel resources available for execution. This is reflected in issues like process creation and deletion, process switching, load balancing and process migration, data copying or sharing, memory management and garbage collection, and synchronization.

Evocation and control of parallelism can be handled implicitly by specific tools, the compiler, or the operating system, or explicitly under the user's control. The first approach certainly is more comfortable for the user, but satisfying tools for parallelization are extremely difficult to develop, and basically not available yet. Explicit user control on the other hand is indispensable for good efficiency, but usually difficult to handle and requires good knowledge of the particular computer system. In the case of logic programming, most explicitly parallel approaches require specific annotations, are totally dependent on a particular evaluation model and execution mechanism, and, in addition, have questionable logical semantics. This, however, is not necessarily a flaw of logic in general; the meta-evaluation concept developed in Chapters 5 and 6 of this book, and related approaches (Shapiro, 1986; Silver, 1986; Goguen and Meseguer, 1987; Goto et al., 1988) present a uniform logic-based methodology, suitable both for natural problem specification and efficient program execution; for a short overview of meta-programming, see (Lloyd, 1989). A problem may be specified in a purely declarative way first, then an adequate evaluation mechanism may be constructed, and execution can be controlled as far as necessary; a change in the execution control or a modification of the evaluation mechanism does not affect the original specification. Knowledge of system details is only necessary if efficiency is very essential, and can be introduced to speed up an existing solution.

Restrictions to the full potential of parallelism will always be present because of reasons inherent to the problem, data dependencies, features of the evaluation paradigm (computational model), properties of the execution mechanism, limited resources, or economy (too much overhead, not worth while). The combination of logic and parallelism, however, has also the capacity to open new classes of problems to computational treatment, extend toy problems to the real world, and move applications from academic to commercial usage.

Books describing aspects of the combination of logic and parallelism – mainly via parallelizing PROLOG – are (Shapiro, 1988; Foster, 1990; Taylor, 1989; Foster and Taylor, 1989). *ACM Computing Surveys* features an overview article on concurrent logic programming languages (Shapiro, 1989), and further recent developments can be found in journals like *Future Generations Computer Systems*, *New Generation Computing*, or in the conferences on *Fifth Generation Computer Systems*, as well as in the sources for AI and parallel computer systems in general.

## 1.4   Relevance of the Work

The work described in this book propagates the combination of logic and parallelism in order to achieve significant mutual benefits. It does not provide a concrete formalism, nor an implementation, but is aimed at an investigation of the potential inherent in a combination of these two areas; it examines important aspects of parallelism to be exploited in logic, as well as the means of logic to describe and govern parallelism.

## Scientific Significance

Focusing on logic as underlying programming methodology, this book is based on the assumption that parallelism allows for an enhancement of sequential processing in a number of directions:

- efficiency,
- natural description, and
- non-determinism.

The contribution of logic is to provide a framework for a parallel programming language and design methodology with features like

- good expressiveness,
- sound semantics and theoretical background,
- high-level declarative programming style,

together with proper integration of execution control for efficiency.

The main results of the book lie in the identification and discussion of categories of parallelism far beyond the ones typically associated with parallelizations of PROLOG, as well as in the derivation of extensions for a logic-based language suitable both for the specification of parallel problems and the parallel evaluation of logic programs. As a consequence, logic can provide the basis for an integrated set of tools for parallel programming, reaching from high-level specification to efficient execution control.

## Scope

The work described in this book touches many important areas in the field of computer science.

- Theoretical Foundations: Logic can be viewed as the core of systems for information processing (Manna and Waldinger, 1985), allowing for a sound investigation and description of important aspects like decidability of problems, correctness of programs, and complexity of algorithms.
- Program Development and Methodologies: It proposes the use of a language based on logic for the design and development of parallel systems, ranging from specification and implementation of programs over the description of parallel architectures to the verification of parallel systems.

- Language Design & Implementation: Based on an existing language without parallel features, a language for the specification and implementation of parallel systems is outlined together with an evaluation mechanism suitable for implementation on parallel systems.

- Architecture: Based on a computational model for the execution of logic programs in parallel, a architecture is developed which allows for the exploitation of parallelism on levels far beyond the traditionally envisaged AND-/OR-parallelism.

- Applications: The development of an inference mechanism more powerful and flexible than existing ones, but with similar or better efficiency, opens new roads to applications for logic programming. At the same time, a mechanism for the specification, implementation and verification of parallel and distributed systems becomes available, with the potential of overcoming some of the hindrances for the appropriate use of such systems.

Obviously not all of these aspects are covered in detail by the book; the contributions made, however, support the belief that logic can provide an adequate means to master some of the challenges imposed by parallelism.

The proposed model offers a basis for a straightforward design and development methodology in the domains of logic programming and parallel system design. It enhances logic programming by a proper integration of modularity, which also serves as a sound basis for the description of parallel systems on various levels of abstraction. Parallel programming is a promising area in the use of computer systems, but seriously suffers from the lack of appropriate tools for easy usability. A unified framework for the design, realization and use of parallel systems contributes to the applicability and acceptance of this technology in many domains. Parallel systems have the potential to provide the high-end computing power which at the moment is almost exclusively available through extremely expensive supercomputers. In combination with an appropriate methodology for the application of parallel programming techniques (as proposed in this book), the economical use of extremely powerful computing systems becomes open to new, financially weaker user groups such as universities, research institutions and smaller companies. The design and realization of parallel systems requires different ways of describing and analyzing problems, and in many cases this may be more appropriate to the nature of problems since it does not require an artificial sequentialization. In the long term, it leads to an withdrawal from the technique of treating small, isolated problems in favor of investigating complex, interconnected systems. The use and acceptance of parallel systems heavily depends on the ease of their usability; this, in turn, is advanced by a consistent and comprehensive methodology for the design and implementation of parallel applications.

## 1.5  Organization of the Book

After some introductory remarks elucidating the background of logic and parallelism mainly by references to established textbooks and publications, the relevance of the work described in this book is outlined, and an overview of its organizational structure is given in the first chapter.

In the second chapter, the *foundations* for the work presented in this book are briefly resumed. From the logic side, these are the Connection Method as the underlying calculus, together with its more specialized variant *model elimination*, and the language LOP. The first two serve as basis for the investigation of the different kinds of parallelism found in Chapter 4; model elimination in addition delivers the computational model for PARTHEO, a *par*allel *theo*rem prover developed within the European ESPRIT project 415 F (Ertel et al., 1989; Bayerl et al., 1989; Schumann and Letz, 1989)[2]. The language LOP (Kurfeß et al., 1989) is a logic programming language with better expressiveness than PROLOG while maintaining clean semantics. The parallel specification language UNITY (Chandy and Misra, 1988) is used to describe the concepts extracted from the investigation of parallelism in logic; since it is already applied to the description of model elimination, it is introduced directly after the section on the Connection Method (Bibel, 1987), featuring a propositional logic variant thereof as introductory example.

The third chapter resumes the state of the art in the area of logic and parallelism. The main aspects of systems combining logic and parallelism are discussed on the levels of the *language* used for pro-

---

[2] the details of PARTHEO itself, however, are not covered here since it is the subject of a Ph. D. thesis by J. Schumann at Technical University, Munich (Schumann, 1991)

gramming, the *computational model* applied for an evaluation of the language, the *abstract machine* to execute the code delivered during the evaluation, and the actual *hardware* to implement the system. Short characterizations of the main projects are given in the form of overview tables.

Chapter 4 features an investigation of various kinds of parallelism relevant for logic programming and the evaluation of logic programs. After a brief overview on the variations of parallelism in logic, each *category* of parallelism is analyzed thoroughly, starting from coarse over medium to small grain. The analysis of a category starts with an informal description of the problem, identifying the important aspects thereof. Then the design of a solution is outlined, resulting in a UNITY program formally specifying this solution. This formal specification is the basis for considerations about the correctness and complexity of the solution proposed, as well as for an discussion of its inherent potential for parallelism. This analysis may lead to refinements of the solution proposed. In addition to the isolated analysis of a particular category, its potential concerning both combinations attained by multiple instances of the category, as well as with previously presented categories is examined.

The insight gathered during these investigations resulted in the outline of a *parallel logic language*, which is described in the fifth chapter. MMLOP stands for *modular meta* LOP and consists of an enhancement of LOP with the concepts of *modularity* and *meta-evaluation*. LOP, extended with these two concepts, is considered sufficient for both the *description* of parallel systems and the efficient *evaluation* of (parallel) logic programs on parallel computer systems.

In the sixth chapter, a computational model for the evaluation of MMLOP programs is described. It is based on one hand on (global and local) *meta-evaluation* aiming at efficiently executable code; on the other hand, it applies algebraic transformations on the modules of a MMLOP program in order to achieve a good correspondence between the structure of the problem and the topology of the underlying architecture.

Implications of a particular computational model called *spanning setters*[3] derived from the Connection Method, on three classes of hard-

---

[3] this term is coined after *spanning sets*, an important concept in the Connection Method

ware architectures are discussed in Chapter 7. Due to the importance of unification in general, and its potential for parallelism in the spanning setter approach in particular, the evacuation of unification onto dedicated coprocessors is also examined in that chapter.

Although in many aspects only the surface could be scratched, some conclusions on the suitability of logic as a tool for parallel programming can be drawn in the last chapter; the amount of work to be done in the future, however, is at least as great as the quality of the perspectives visible.

# Chapter 2

# Foundations

The following chapter provides the foundations for the work described in this book. The main part of the chapter concentrates on the logical foundations, starting with an outline of essential features of *formal logic*. Then the *connection method* is described, a logical calculus with some interesting properties, especially with respect to parallelism.

Important concepts for the *specification of parallelism* are described based on the parallel specification language UNITY; as an introductory example, a variant of the connection method for propositional logic is specified with UNITY. The concepts required for parallel *execution* of programs rely on processes as basic mechanisms.

A variant of the connection method, namely *model elimination*, serves as evaluation mechanism for a family of theorem provers (SETHEO, PARTHEO) with a close relation to the work presented in this book. Therefore, a detailed specification of model elimination in UNITY is given. LOP, the input language for SETHEO and PARTHEO concludes the chapter on foundations.

# 2.1   Logic

The precise formulation of a problem to be solved is a very important step towards its successful solution. For such a *problem specification* a notation which guarantees the non-ambiguity of the statements made is indispensable. The same basically holds for the description of the way to achieve a solution,[1] especially if the problem is to be treated by a computer. Conventional programming languages are very well suited for formulating the solution process for a class of problems which can be solved efficiently by the application of an *algorithm*; however, there might be a considerable gap between the abstract problem specification and the actual program implementing the solution algorithm. Programming paradigms based on the use of mathematical logic as specification and implementation language[2] offer a tool for bridging this gap, thus allowing a more direct derivation of the actual program from the formal specification. The logic approach, however, still suffers from two major hindrances to its broad acceptance:

**Efficiency:** The evaluation of logic programs on conventional computers is quite slow compared to procedural languages. This is caused mainly by two reasons. First, the evaluation process is basically a *search* of the space of solutions for (the) one that fulfills the conditions given by the program. Second, the underlying computation mechanism for the exchange of values between different parts of the program is *unification*; this is considerably more difficult to implement on conventional machines than the assignment operation, which is the equivalent method for transferring values in procedural languages.

**Expressiveness:** The expressive power to describe certain features of problems in practice is somehow limited, especially for logic programming tools with reasonable efficiency (e.g. PROLOG). It is difficult, for example, to describe uncertain, inconsistent or incomplete knowledge (the exceptions to the general rules), or to include time (temporal relations). In principle, these deficiencies can be overcome by purely logical means, but with considerable loss of efficiency. On the other hand, the inclusion of non-logical features provides an escape, this time giving up a certain degree of "cleanliness". An

---

[1] such a description may contain a certain degree of non-determinism
[2] PROLOG is the most popular of these "logic" programming languages

alternative might be to combine logic with other formalisms suited for the treatment of this kind of knowledge (Zadeh, 1983; Shafer and Pearl, 1990).

In summary, (mathematical) logic provides a formal tool for the derivation of conclusions, using symbols as objects of computation. For the discussion of a system based on logic, two important issues have to be considered:

**Syntax,** which defines the language used to express statements in logical terms. It determines the composition of legitimate expressions, or *well-formed formulae*, by giving the grammar of a formal language including the deduction rules.

**Semantics,** which aims at establishing a correspondence between the elements of the language and the entities, relations, and functions in the domain of discourse. It involves the assignment of *meaning* to formulas of the formal language by providing an *interpretation* of a formula; its *evaluation* yields **true** or **false**, respectively.

A *calculus* is a set of rules, describing a method of deriving new logical formulae from given ones; these new formulae are called *theorems*, and the sequence of rule applications during a derivation of a theorem is a *proof* for the theorem. The method of deriving theorems within a logical calculus is based on

**Axioms,** which represent the basic assumptions, and

**Inference rules,** which specify the cases where a new formula can be derived from already available ones.

Important properties of a calculus are

**Correctness:** Only *semantically* valid formulae can be deduced *mechanically*; a correct calculus is also called *sound*, and

**Completeness:** Each *semantically* valid formula can also be deduced *mechanically*.

In the case of logic programming, another mandatory feature is the efficiency that a calculus offers for the implementation on a computer; if it is not sufficient, the calculus will not be used very much.


## Propositional Logic

The logical *combination of statements*, or *propositions* is the object of study in propositional logic. The internal structure of such a proposition is completely disregarded, and thus a whole proposition (although it can be a quite complex statement) may be abbreviated by a single symbol.

### Syntax

The language of propositional logic comprises *sentences* or *formulae*, composed of *propositional symbols* or *propositional variables* and *connectives* or *operators* according to some syntactical rules. It deals with propositional symbols only (which can be regarded as nullary predicates), and it does not contain constants, variables, terms, functions, or quantifiers.

### Semantics

A *meaning* is given to such a sentence or formula of propositional logic by assigning a truth value to it. An *interpretation* $\mathcal{I}$ for a formula $\mathcal{F}$ is an assignment of a truth value to each of the occurring propositional symbols; all occurrences of one particular symbol must have the same value.

### Properties of Sentences

Important basic properties of sentences in propositional logic are

**Satisfiability:** It is possible to find an interpretation $\mathcal{I}$ for a formula $\mathcal{F}$ such that $\mathcal{F}$ is true, and

**Validity:** $\mathcal{F}$ is true for every interpretation $\mathcal{I}$.

A sentence $\mathcal{F}$ is *contradictory* or *unsatisfiable* if there is no interpretation $\mathcal{I}$ such that $\mathcal{F}$ is true.

### Computational Evaluation

The truth value of a particular formula can be determined by applying *semantic rules* according to the operators defining the structure of the formula. A satisfying way of evaluating smaller sentences is given by *truth tables.* For larger sentences, the inspection of the corresponding *semantic tree* usually is better suited. In principle, a complete analysis of all possible truth value assignments can be made to decide if a formula is valid or not; efficiency reasons, however, usually ask for more sophisticated methods.

## Predicate Logic

The expressive power of propositional logic is quite restricted; it allows only for the combination of basic statements, without taking into account their internal structure. In *(first order) predicate logic*, more sophisticated problems can be dealt with by the introduction of some additional concepts: *Constants* and *variables* are used as symbols to denote objects; *predicates* express properties of or relations between objects; *functions* describe operations to be performed on objects, and *quantifiers* can be seen as generalizations of the connectives in propositional logic. Constants, variables and functions can be composed into *terms*, thus allowing for arbitrarily complex object descriptions.

### Syntax

A typical notational variant used in logic programming is *clause form*; a formula is grouped into *clauses* consisting of a *head* and a *body*, basically representing an implication from the body to the head. Quantifiers do not appear explicitly; each clause, however, is implicitly assumed to be universally (or all-) quantified.

### Semantics

As in propositional logic, meaning is given to a formula by an *interpretation*. However, the case of predicate logic is more complicated; it is not sufficient to just assign truth values to the symbols since the formulae may contain arbitrarily complex term structures. Thus, an interpretation $\mathcal{I}$ is based on a *domain* $\mathcal{D}$, and the set of objects of $\mathcal{D}$ together with functions and relations over $\mathcal{D}$ provide a meaning for constant, (free) variable, function and predicate symbols. With such an interpretation $\mathcal{I}$ we can determine the value of a formula $\mathcal{F}$ (which is either **true** or **false**) or of a term $\mathcal{T}$ (which must be an object in the domain $\mathcal{D}$). The value of a complicated expression is determined stepwise by applying the appropriate semantical rules.

### Properties of Sentences

The most important properties of sentences are basically the same as in propositional logic:

**Satisfiability:** An interpretation $\mathcal{I}$ for a formula $\mathcal{F}$ can be found such that $\mathcal{F}$ is true;

**Validity:** $\mathcal{F}$ is true for any interpretation $\mathcal{I}$.

A sentence $\mathcal{F}$ is *contradictory* or *unsatisfiable* if there is no interpretation $\mathcal{I}$ such that $\mathcal{F}$ is true.

## Computational Evaluation

In contrast to propositional logic, the determination of the validity of a predicate logic formula is generally not possible via a complete analysis of all possible combinations. More sophisticated proof procedures have to be applied whose efficiency largely rely on the underlying calculus. For the computational treatment of predicate logic, two classes of calculi are of high interest:

**Full First Order Predicate Logic,** which is correct and complete, but inefficient; the corresponding semantics is simple and elegant, but not very well suited for computation. There is no minimal Herbrand model which could be viewed as the semantics of a logic program.

**Horn Clause Logic** is a *subset* of full first order predicate logic with only one literal in the clause head. It has a minimal Herbrand model whose model-theoretic semantics is equivalent to the fixed-point and operational semantics; negation, however, is only available "as failure". In addition, PROLOG exists as reasonably efficient implementation, but has problems with correctness (occur check), completeness (search strategy), and the inclusion of non-logical features with questionable semantics.

In general, logic is a very powerful and, for a wide class of problems, well suited programming paradigm; however, its application still suffers from relatively high execution times.

## Limitations of Formal Logic

One of the typical properties of a formal logic system is the clear separation between the framework for representing information, the *language*, and the mechanism for drawing conclusions from these information items, the *inference rules*. This separation gives immediate rise to two fundamental problems: First, a language has to be found that has the expressive power to capture relevant information in an appropriate way; neither PROLOG nor different "pure" logic approaches (e.g. predicate, modal, temporal logic) are fully satisfying for the description of so-called "real-world" phenomena. Second, an inference mechanism reasonably close to human reasoning should be available or at least easily recomprehensible. Criticizers of the logic (or logistic, as some prefer to call it) approach claim that these flaws are fundamental to formal logic, and independent of the particular formalisms (Minsky, 1981). Some of the objective points are

**Simplicity:** Only microworlds of limited complication, composed of many small items, can be described.

**Size:** The amount of data items within such a model is too small to comprise realistical problems, or the problem becomes intractable due to combinatorial explosion.

**Relevance:** It is difficult to extract relevant issues from a variety of descriptive data items; in addition, relevance can be subjective and arbitrary, i.e. based on individual judgement.

**Focusing:** There is a danger that a model is trimmed in the sense that only those issues are included into the description which permit one to derive wanted conclusions and exclude unwanted ones.

**Formalization:** The contents and structure of knowledge are very often only vaguely accessible, and the construction of an appropriate formal system leads to intolerable simplifications; problems especially arise with concepts like time, locality, procedural knowledge, cause-and-effect, purpose and similar ones.

**Control:** The deduction process in general is encapsulated and not accessible to the user; this has to be achieved by non-logical means, or the introduction of an extra control level (meta-level).

**Consistency (Correctness):** A logical model of a system is usually aimed at a consistent description without internal contradictions, whereas in reality inconsistencies just exist; an important aspect of intelligence, however, is the ability to deal with inconsistencies and contradictions.

**Completeness:** All true statements can be proven by the system; this feature can also be achieved by exhaustive search, and does not necessarily require a logic formalism.

**Monotonicity:** In classical logic systems, the addition of further axioms must not influence the validity of already derived conclusions, leading to the construction of self-consistent models.

Obviously some of these objective points are equally well applicable to formalisms or methods other than logic, and might be more inherent to formal systems in general than to logic in particular. The latter three points especially have a certain fascinating mathematical virtue and elegance, and surely are indispensable for the use of logic as a *proof mechanism*; however, it is questionable whether they must be incorporated into common-sense reasoning systems. And on the other hand, logicians are still optimistic about coming up with mechanisms that are better than the existing ones.


## Connection Method

The connection method is the basic framework for a family of complete and sound calculi for first order logic (Bibel, 1987; Bibel, 1983). Its central evaluation paradigm consists of checking paths through specific parts of the formula for tautology in the case of a direct proof (or for contradiction in a refutation proof). To ease visualization, a formula is represented as a matrix; an example of a formula in clausal form together with its matrix notation is given in Figure 2.1. The clauses of the formula correspond to the columns of the matrix, and the literals of the clauses to the elements of the columns. A *path* through a matrix is a set of literals, one from each column of the matrix. The criterion for the *validity of a formula* is that all possible paths through the corresponding matrix are *complementary*, i.e. they contain a literal together with its negation (which is called a *connection*), and compatible substitutions for the variables occurring in these complementary pairs of literals can be found (the connections must be *unifiable*). Checking all possible paths of a matrix of course is very tedious, and, in general, not necessary; a proof can already be provided by a subset of the existing literals and their connections, which must fulfill a certain criterion ("spanning set"). In addition, alternative solutions might exist, corresponding to multiple "spanning sets" of connections. In the extreme case, such a check of all paths could result in an exponential amount of work to be performed; thus, an implementation aimed at efficiency would use some optimizations reducing the search space to be traversed drastically.

After this very short résumé we describe the most important concepts of the connection method in some more detail; we restrict ourselves, however, to formulae in clause form, represented by matrices. An in-depth treatment of arbitrary formulae, as well as proofs for the theorems given, can be found in (Bibel, 1987).

$$\underbrace{\neg P(a)}_{\text{clause 1}} \;\lor\; \underbrace{(P(x) \land \neg P(fx) \land Q(x))}_{\text{clause 2}} \;\lor\; \underbrace{P(ffa)}_{\text{clause 3}} \;\lor\; \underbrace{(R(y) \land \neg Q(y)}_{\text{clause 4}} \;\lor\; \underbrace{\neg R(z)}_{\text{clause 5}}$$

| clause 1 | clause 2 | clause 3 | clause 4 | clause 5 |
|---|---|---|---|---|
| $\neg P(a)$ | $P(x)$ | $P(ffa)$ | $R(y)$ | $\neg R(z)$ |
|  | $\neg P(fx)$ |  | $\neg Q(y)$ |  |
|  | $Q(x)$ |  |  |  |

Figure 2.1: A predicate logic formula and its matrix notation

**Propositional Logic**

A propositional logic formula is composed of symbols connected by logical operators; parentheses may be used for structuring.[3]

**Definition 2.1 (Symbols)**

The symbols used in a propositional formula are the *truth symbols* **true** and **false**, and *propositional variables*, denoted by capital letters, typically $P, Q, R$. ∎

Propositional variables can stand for one of the truth values, as well as for whole (sub-)formulae, or *propositions*.

**Definition 2.2 (Literal)**

A (positive or negative) *literal* is an un-negated (positive) or negated (negative) propositional variable. ∎

Usually only negated literals are explicitly marked by a preceding negation sign ($\neg$).

**Definition 2.3 (Complementary Literals)**

Two literals are called *complementary* if they have the same propositional variable, but with different sign. ∎

**Definition 2.4 (Clause)**

A *clause* is a conjunction of literals; a *unit clause* contains only one literal. ∎

**Definition 2.5 (Matrix)**

A *matrix* is a disjunction of clauses; the clauses are represented vertically as columns, and the disjunction and conjunction operators ($\lor$ and $\land$) are omitted. ∎

Such a matrix is a representational variant for a formula, which is more convenient for visualizing the connection method than the standard

---

[3] in most cases we use a normal form where structuring is not necessary (usually disjunctive normal form, or clause form)

notation. The above definition of a matrix implicitly is based on a normal form, namely clausal form; this is not a necessity, but just for the sake of convenience. As an example, let's have a look at the following formula:

$$\Gamma(A \vee B) \wedge (A \wedge C \to B) \wedge C \to B$$

In clausal form, the formula looks as follows:

$$(\neg A \wedge \neg B) \quad \vee \quad (A \wedge C \wedge \neg B) \quad \vee \quad \neg C \quad \vee \quad B$$

clause 1              clause 2              clause 3        clause 4

Finally, its matrix notation simply is a spatial rearrangement of the clausal form:

$$(\neg A \wedge \neg B) \quad \vee \quad (A \wedge C \wedge \neg B) \quad \vee \quad \neg C \quad \vee \quad B$$

| clause 1 | clause 2 | clause 3 | clause 4 |
|----------|----------|----------|----------|
| $\neg A$ | $A$ | $\neg C$ | $B$ |
| $\neg B$ | $C$ | | |
| | $\neg B$ | | |

Finding a proof for a certain formula is based on searching all paths through the matrix representing the formula.

**Definition 2.6 (Path)**

A *path* through a matrix consists of a set of literals, precisely one from each clause of the matrix.                                          ∎

The criterion for which the paths of a matrix are searched is

**Definition 2.7 (Complementary Path)**

A path is *complementary* iff[4] it contains at least two complementary literals.                                          ∎

**Definition 2.8 (Connection)**

A pair of complementary literals in a path is called a *connection*.   ∎

Based on the notions of complementarity and connection, a criterion for determining the validity of a propositional logic formula (in clause form) can be given:

**Theorem 2.9 (Validity: Complementary Paths)**

A matrix represents a valid formula iff each path through the matrix is complementary, i.e. each path contains a connection.                     ∎

This theorem is based on a check of *all* paths through the matrix; in general, this is not necessary. It is sufficient to check a particular

---

[4]"iff" is used as shorthand for "if, and only if"

subset of all paths, representative for the underlying structure of the formula. Due to historical reason, sit is also common to use the *negative representation* of a formula $\mathcal{F}$, which is simply its negation $\neg\mathcal{F}$; it can be shown that $\mathcal{F}$ is valid iff $\neg\mathcal{F}$ is *unsatisfiable*, i.e. there is no interpretation $\mathcal{I}$ such that $\mathcal{F}$ is true.

**Definition 2.10 (Spanning Sets)**

A set $\mathcal{S}$ of connections is called *spanning* for a matrix $\mathcal{M}$ if each path through $\mathcal{M}$ contains at least one connection from $\mathcal{S}$. ∎

Based on this characterization of a spanning set, the criterion for the validity of a formula can be rephrased as follows:

**Theorem 2.11 (Validity: Spanning Sets)**

A matrix represents a valid formula iff there exists a spanning set of connections for it. ∎

If more than one spanning sets can be identified for a formula, they stand for alternative solutions (proofs).

An application of these concepts to the above example yields

$$
\begin{array}{llll}
\{\neg A & A & \neg C & B\} \\
\{\neg A & C & \neg C & B\} \\
\{\neg A & \neg B & \neg C & B\}
\end{array}
\qquad
\begin{array}{llll}
\{\neg B & A & \neg C & B\} \\
\{\neg B & C & \neg C & B\} \\
\{\neg B & \neg B & \neg C & B\}
\end{array}
$$

as all possible paths through the formula; as can be seen easily, each of the paths contains a pair of complementary literals, or a *connection*, and the formula is valid.

The plausibility of this method may become clearer if a path is interpreted as a disjunction of literals, and the whole formula as a conjunction of paths. Then the whole formula is valid if all paths are tautological; this is exactly the case if each path contains a pair of complementary literals.

## Predicate Logic

Since the expressive power of propositional logic is quite restricted, a system for logic programming has to be able to deal with a more powerful calculus; an obvious candidate here is first order predicate logic[5]. In the following, the concepts introduced above are – as far as

---

[5] in contrast to many logic programming systems restricted to Horn clause logic (e.g. PROLOG), the connection method deals with *full* first order predicate logic

possible – 'lifted' from propositional to first order logic; in addition, a few new ones have to be introduced.

A first order predicate logic formula may comprise representatives from five classes of symbols; in addition to the logical connectives from propositional logic, quantifiers may occur. In this context, where only formulae in clausal form are regarded, quantifiers as well as most logical connectives do not appear explicitly.

## Definition 2.12 (Symbols)

The symbols used in a first order formula are:

- the *truth symbols* **true** and **false**,
- *constants*, denoted by lower case letters or strings, typically $a, b, c$,
- *variable symbols*, denoted by lower case letters or strings (starting with a lower case letter), typically $u, v, w, x, y, z$,
- *function symbols*, again denoted by lower case letters or strings, typically $f, g, h$,
- *predicate symbols*, denoted by lower case names, typically $P, Q, R$.[6]

∎

Intuitively, constants and variables denote objects, whereas function and predicate symbols stand for functions and relations, respectively. Function and predicate symbols have an associated *arity*, indicating the number of arguments for the corresponding function or relation. Note that *propositional variables* do not appear in predicate logic. In predicate logic, objects can be described not only by constants or variables, but also by more complicated expressions called terms, which can be used to specify objects in more detail.

## Definition 2.13 (Term)

The *terms* of predicate logic are composed according to the following rules:

- *Constants* are terms.
- *Variables* are terms.
- If $f$ is a *function symbol* of arity $n \geq 1$, and $t_1, t_2, \ldots, t_n$ are terms, then the *application*

$$f(t_1, t_2, \ldots, t_n)$$

is a term.

∎

Based on the concept of terms, the definition of a literal can now be

---

[6] this notation is common for mathematical logic; PROLOG unfortunately denotes variables by capital and predicates by lower case letters or strings

extended to predicate logic.

**Definition 2.14 (Literal)**

A (positive or negative) *literal* is constructed according to the following rules:

- **true** and **false** are literals.

- If $P$ is a *predicate symbol* of arity $n \geq 1$, and $t_1, t_2, \ldots, t_n$ are terms, then

$$P(t_1, t_2, \ldots, t_n)$$

   is a literal.

- If $L$ is a literal, then also $\neg L$.

Again, only negated literals are explicitly marked by a preceding negation sign $(\neg)$.

**Definition 2.15 (Complementary Literals)**

Two literals are called *complementary* if they have the same predicate symbol, but with different sign.

The composition of clauses is identical to propositional logic:

**Definition 2.16 (Clause)**

A *clause* is a conjunction of literals.

A formula consisting of a set of clauses is implicitly existentially quantified; this has the consequence that identical variable symbols appearing in different clauses have different binding scopes (the respective clause), and thus may have different values. Each clause, in contrast, is universally quantified, and identical variable symbols in the same clause must have the same value; they are called *shared* variables with respect to the literals of the clause.

The definition of a matrix again is identical to propositional logic:

**Definition 2.17 (Matrix)**

A *matrix* is a disjunction of clauses; the clauses are represented vertically as columns, and the disjunction and conjunction operators ($\wedge$ and $\vee$) are not explicitly represented.

As in propositional logic, a proof attempt is based on checking all paths through the matrix representing the formula.

**Definition 2.18 (Path)**

A *path* through a matrix consists of a set of literals, precisely one from each clause of the matrix.

The criterion for which the paths of a matrix are checked is again complementarity.

**Definition 2.19 (Complementarity)**

A pair of literals is *complementary* iff it the predicate symbols of the two literals are identical, but with different sign (one positive, one negative). ▪

**Definition 2.20 (Connection)**

A pair of complementary literals is called a *connection*. ▪

**Definition 2.21 (Spanning Set)**

A set $S$ of connections is called *spanning* for a matrix $M$ if each path through $M$ contains at least one connection from $S$. ▪

In contrast to propositional logic, the existence of a spanning set is not sufficient for the validity of a predicate logic formula: the structure of the terms appearing as arguments of the predicates in the literals of a connection must be compatible, and 'consistent' variable bindings must exist. This is expressed in the concept of unification, based on substitutions for variables (Robinson, 1965; Eder, 1985; Lassez et al., 1986).

**Definition 2.22 (Substitution)**

A *substitution* $\sigma = \{v_1 \leftarrow t_1, v_2 \leftarrow t_2, \ldots, v_n \leftarrow t_n, \}$, $v_i \neq t_i$, assigns terms to variables; it is applied to a term $t$ (denoted by $\sigma(t)$, $\sigma t$, or $t\sigma$) by simultaneously replacing in $t$ all occurrences of each $v_i$ by the corresponding $t_i$. ▪

The process of finding a substitution such that variable bindings are consistent is described by

**Definition 2.23 (Unification)**

Two terms are *unifiable* if there is a substitution for the occurring variables which makes the two terms identical; or, more formally and extended to sets: For a set $T$ of terms $t_i$ a substitution $\sigma$ is called a *unifier* of $T$ iff $\sigma(t_i) = \sigma(t_j)$ for all $t_i, t_j \in T$. ▪

**Definition 2.24 (Most General Unifier)**

A unifier for a pair of terms is called *most general unifier* or *mgu*, for short, if each other unifier can be obtained from the mgu through specialization; the mgu maps two terms to their most general common instance. ▪

A criterion for determining the validity of a predicate logic formula (in clause form) can be given based on the notion of spanning sets of connections and unification:

**Theorem 2.25 (Validity: Unifiable Spanning Set)**

A matrix represents a valid formula iff there exists at least one *unifiable* spanning set of connections for it.                                    ∎

If more than one spanning sets can be identified for a formula, they stand for alternative solutions (proofs). As in propositional logic, a proof of a formula $\mathcal{F}$ can also be achieved by showing that its negation $\neg\mathcal{F}$ is unsatisfiable.

With the example from the beginning of this section (**Figure 2.1**), we obtain the following:

a) the matrix with the connections:

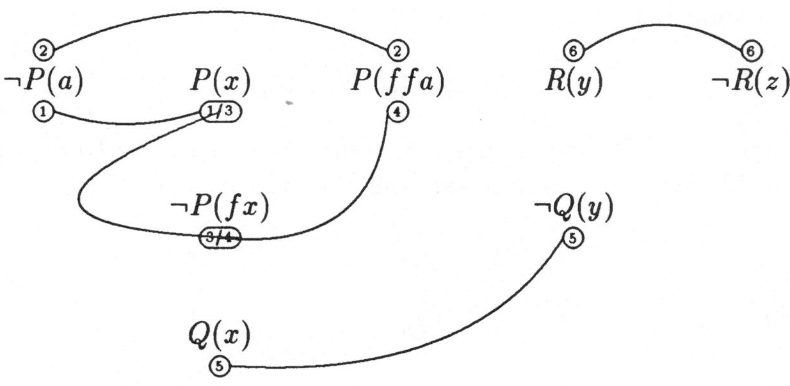

Figure 2.2: The matrix with connections

b) the spanning sets of connections are $\{2\}$ and $\{1,4,5,6\}$; the first spanning set is not unifiable ($a \neq ffa$), and for the second unification fails as well since $a = x$ and $fx = ffa$ leads to the contradiction $fa = ffa$; this contradiction, however, can be resolved using connection 3, which represents a cycle.[7] A copy of the clause containing this connection is made, and as a consequence copies of other clauses are required.

c) the matrix with copies (see Figure 2.3). Now, the expanded spanning

---

[7] cycles basically represent recursive parts of a formula; see (Bibel, 1988) for a more detailed discussion

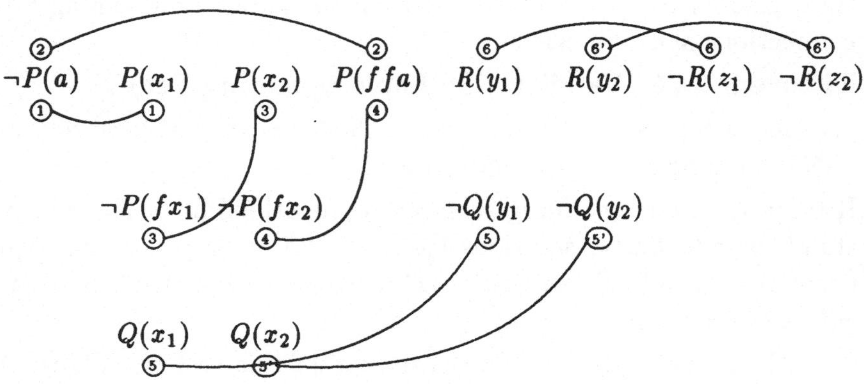

Figure 2.3: The matrix with copies

set comprises the connections

$$\{1, 3, 4, 5, 5', 6, 6'\}$$

where $5'$ and $6'$ denote copies of connections 5 and 6. Unification is possible for this spanning set, and the substitutions

$$
\begin{array}{ccccccc}
x_1 & \leftarrow & y_1 & \leftarrow & z_1 & \leftarrow & a \\
x_2 & \leftarrow & y_2 & \leftarrow & z_2 & \leftarrow & fx_1 \\
fx_2 & \leftarrow & ffa
\end{array}
$$

validate the formula.

Please note that the copies required for the proof of the formula do not have to be represented explicitly; it is sufficient to introduce distinct instances of the variables involved. The explicit representation given here is merely used to ease visualization.

A concept not required right now but useful in later sections is a restriction of unification to the structure of the terms involved, without regarding variable bindings induced by other unifications (Eder, 1985):

**Definition 2.26 (Weak Unification)**

Two terms $t_1, t_2$ are *weakly unifiable* iff there is a pair of substitutions $\sigma_1, \sigma_2$ such that $\sigma_1(t_1) = \sigma_2(t_2)$.                               ∎

Weak unification requires relatively little effort compared to full unification; it can be applied favorably to confine the selection of potential unification partners for a literal.

**Theorem 2.27 (Validity: Disjunction of Literals)**

A disjunction of literals existentially quantified is valid iff it contains
a weakly unifiable complementary pair of literals.                      ∎

Another proposition useful during the evaluation process of a formula
concerns the relation between propositional and predicate logic via the
abstraction of a predicate logic formula to its propositional 'skeleton':

**Theorem 2.28 (Relation Propositional – Predicate Logic)**

If a predicate logic formula is valid, then also its *propositional abstraction*, given by using the predicate symbols as propositional variables
and skipping the rest.                                                  ∎

The opposite direction obviously does not hold, since in the predicate
logic case, unification might fail. For a brief check of the *invalidity* of
a predicate logic formula, the negation of the above theorem can be
used, saying that if the propositional restriction of a predicate logic
formula is invalid, then also the predicate logic formula itself.


## 2.2   Parallelism

Describing problems with inherently parallel properties as well as the
execution of programs on parallel computer systems poses strong demands on the development of programs, starting form the original
problem identification, over the design, specification and implementation of a solution, to the actual program execution. Maybe even
more than in the sequential case, parallel programming must rely on
a good program development methodology, an appropriate language
framework, and sufficient support for the execution of programs. Unfortunately, the state of the art in this area is not too advanced yet,
and compromises are mandatory. Although this book advocates the
use of logic for all issues mentioned above, it turned out to be unfeasible to describe the exploitation of parallelism in logic, and at the same
time develop a framework based on logic for the description of parallelism. Thus, for specification purposes Unity was chosen among
a number of candidates (e.g. FP2 (Jorrand, 1986), Occam (INMOS,
1988b), or some of the parallel logic languages like GHC (Ueda, 1985),
Concurrent Prolog (Shapiro, 1988), Parlog (Clark, 1985)) mainly
because it not only provides a language framework, but also a development methodology together with a mechanism to reason about crucial

properties of programs and algorithms (like termination, complexity, invariants). For the description of features related to program execution, *processes* were chosen because they combine a sound theoretical background (Hoare, 1985) with considerable practical experience. These two concepts, UNITY and processes, are the main subjects of this section.

## UNITY

The language UNITY has been designed as a tool to develop programs – especially *parallel* ones – systematically for a variety of architectures and applications; for an in-depth description together with a lot of helpful example programs see (Chandy and Misra, 1988). One of the main purposes of this language is the decoupling of *design, specification, coding* and *implementation* phases during the development of programs. The methodology used relies on

- identification of the core of the problem,
- design of a solution to the problem,
- formal specification of the solution,
- derivation of important properties of the solution (correctness, complexity, adequacy, feasibility),
- systematical derivation of programs for specific architectures and applications,
- implementation of a program on a particular architecture.

In addition, an important concept is the stepwise refinement of specifications, adding incrementally more detail to the 'abstract' solution. This methodology helps to separate a number of concerns which are often intermingled in classical programming: a *program* is decoupled from its *implementation*, and the flow of *control* during program execution does not appear in the specification.

UNITY actually represents a computational model and a proof system, and the phrase "a UNITY program" stands for "a program in *U*nbounded *N*ondeterministic *I*terative *T*ransformation notation" (Chandy and Misra, 1988). Thus, UNITY is not a full programming language, only a notational mechanism to specify certain aspects of programs. Important features of the underlying computational model are *non-determinism* and *state transitions*. Non-determinism, especially in parallel systems, is necessary to express inherently nondeterministic properties of a problem, and often helpful to keep programs simple by avoiding unnecessary determinism. State transition systems are a formal model with a sound theoretical background, and have been successfully applied to many applications, especially modeling of physical systems.

## UNITY **Programs**

A UNITY program consists of a declaration of variables, a specification of their initial values, and a set of multiple-assignment statements[8]. A program is executed – starting from a state satisfying the initial conditions – by selecting nondeterministically one applicable assignment statement and performing the corresponding state transition; this cycle is repeated until there is no applicable statement left. Execution is assumed to proceed forever and is subject to *fairness*, which means that every statement – provided that it is applicable – is selected infinitely often. A program which runs forever does not terminate; termination, however, often is a crucial aspect of a program. UNITY regards termination as an aspect of implementation, and separates the execution of a UNITY program (an infinite sequence of statement executions) from an implementation, which describes a finite prefix of the sequence. Questions related to termination are handled via *fixed points*, which are states of the program where the execution of any statement of the program leaves the current state unchanged.

As an introductory example, we develop a UNITY program which specifies the evaluation of a (propositional) logic formula according to the connection method (described in Section 2.1).

### Problem Specification

For a propositional logic formula $F$, try to find a proof $P$.

### Design of a solution

Our solution is derived from Theorem 2.9, stating that a matrix represents a valid formula iff every path through the matrix is complementary. The strategy is to check every path for complementarity, store this partial result, and combine all the partial results to obtain the overall result. The correctness of our solution is based on the belief

---

[8] the notation used in this book actually is slightly modified with respect to (Chandy and Misra, 1988)

that a logical theorem elsewhere proven to be correct is trustworthy.

The program as shown in Figure 2.4 obviously is not very sophisticated yet; it is derived quite directly from the definitions and theorems used to describe the connection method without much effort. Its purpose anyway is to get some acquaintance with the UNITY notation and the corresponding methodology. In the following, we will discuss the components of the above program in some detail. The first section specifies the names, types and meaning of the *parameters* used; in contrast to *local variables* specified in the third section, parameters are visible to and get their values from the outside world; local variables are relevant and known only inside the program, and should have an initial value. Conditions for the initial state of the program are given in the *pre-conditions* section; in our case, for example, the formula must be given in clause form because otherwise the definition of a path would have to be changed. *Invariants* can serve different purposes; the first is to provide additional assertions about the relations between the values under computation, indicating the correctness of the program. The second is to introduce abbreviations by defining certain program variables as functions of other variables[9]; in our example we used this feature to define complementary paths. The third purpose finally is to specify additional conditions which have to be valid throughout the whole execution of the program; an example is the request that all paths through the matrix are different. Note that the first usage of invariants is somehow passive in the sense that it provides additional safety for the design of the program by specifying conditions which must be fulfilled by the rest of the program; the other two play an active role during the execution of the program and are integral parts of it.

The *computation* section represents the core of a UNITY program, consisting of a set of statements which include assignments describing the state transitions to be performed. Please be aware of the underlying evaluation paradigm of UNITY programs: it is based on a (potentially infinite) cycle, where in each step one applicable statement is chosen nondeterministically and then executed as specified; if a statement consists of multiple assignments, all are executed simultaneously.

In an *assignment*, a value specified by an expression is assigned to

---

[9] it can be seen as kind of a *macro* definition mechanism

**Program COMPLEMENTARY PATHS 1**
**parameters**

| name | type | description |
|------|------|-------------|
| $F$ | logical formula | formula |
| $result$ | {proven, unprovable, undefined} | result of the evaluation |
| $\mathcal{PATHS}$ | set of paths | paths in the matrix |
| $PATH_i$ | set of literals | single path |
| $\mathcal{CLAUSES}$ | set of clauses | clauses of the matrix |
| $CLAUSE_j$ | set of literals | single clause |
| $\mathcal{LITERALS}$ | set of literals | literals of the matrix |
| $L_k$ | propositional variable | single literal |

**pre-conditions**

$F = \{CLAUSE_1, \ldots, CLAUSE_n\}$      clause notation

$result =$ **undefined**      result unknown

**local variables**

$result_i$    {proven, unprovable, undefined}      result of the evaluation

**invariants**

$\forall\, i,j:\ PATH_i \in \mathcal{PATHS} \wedge PATH_j \in \mathcal{PATHS}$   all paths
$\qquad\qquad :: PATH_i \neq PATH_j$   are different

$PATH_i = \bigcup_{1 \leq j \leq\ |\mathcal{CLAUSES}|} L_j : L_j \in CLAUSE_j$   definition path: one literal from each clause

$compl(PATH_i) \equiv$   definition

$\exists\, k,l : L_k \in PATH_i, L_l \in PATH_i :: L_k = \neg L_l$   complementary path

**computation**

$[\!] \quad \langle\ i:\ PATH_i \in \mathcal{PATHS} ::$   for all paths
$\quad [\!]\ result_i :=$ **proven** if $compl(PATH_i)$   path complementary
$\quad [\!]\ result_i :=$ **unprovable** if $\neg compl(PATH_i)$   path not complementary
$\quad \rangle$

$[\!] \quad [\!]\ result\ :=$ **proven** if   formula proven
$\qquad\qquad \forall\, i:\ PATH_i \in \mathcal{PATHS} ::$   if all paths
$\qquad\qquad result_i =$ **proven**   are provable
$\quad [\!]\ result\ :=$ **unprovable** if   formula not provable
$\qquad\qquad \exists\, i:\ PATH_i \in \mathcal{PATHS} ::$   if there are
$\qquad\qquad result_i =$ **unprovable**   unprovable paths

**fixed points**

$\forall\, i:\ PATH_i \in \mathcal{PATHS} :: result_i \neq$ **undefined**   all paths checked

**end COMPLEMENTARY PATHS 1**

Figure 2.4: Program COMPLEMENTARY PATHS 1

a variable, as for example $a := b + c$; expressions may be conditional, indicated by if, but if ... then ... else constructions are not allowed and have to be specified as separate conditional expressions. Very often it is desirable to assign values to different variables within one assignment; UNITY offers three ways to do so:

1. *composed assignment* : the values to be assigned to the different variables in a single assignment are specified in *components* of the assignment, one for each variable; these components are separated by $\|$. The example $x := y \| y := x$ exchanges the values of $x$ and $y$. In many cases, the components of a composed assignment are written in separate lines, such as the two alternative conditional assignments to the variable *result* in the above program; in these cases, each component of the assignment may be preceded by a separator $\|$ for easier readability. A composed assignment may also be more complex, containing enumerated and quantified assignments as components.

2. *enumerated assignment* : the variables to be used in the assignment are listed to the left of the assign operator $:=$, separated by commata; the expressions specifying the corresponding values are given on the right side in the same order. In this notation, $x, y := y, x$ exchanges the values of $x$ and $y$.

3. *quantified assignment* : this form is very convenient to describe operations on arrays or similar structures using indices. A quantified assignment is embraced by $\| \langle \ldots \rangle$ and starts with a quantification condition of the form $variable - list : boolean - expression ::$, followed by the actual assignment (which may be composed again). A quantified assignment denotes zero or more components; these components are made explicit by instantiating the bound variables of the variable list, one component for each instance. In the example

$$\| \langle i : 0 \leq i \leq n :: A_i := A_i * B_i \rangle$$

the explicit version for $n = 2$ would be

$$A_0 := A_0 * B_0 \| A_1 := A_1 * B_1 \| A_2 := A_2 * B_2$$

or, in separate lines,

$$\| A_0 := A_0 * B_0$$
$$\| A_1 := A_1 * B_1$$
$$\| A_2 := A_2 * B_2$$

What has been clarified by now is the structure of (maybe complex) assignments; such an assignment is executed by evaluating the right-hand sides of all its components simultaneously and assigning the resulting values to the variables on the left-hand side.

*Statements* in a UNITY program are used to group correlated assignments and isolate independent ones; statements are separated by the symbol $[\,]$ , and may be quantified in the same way as a quantified assignment, using $[\,]$ instead of $\|$:

$$[\,] \langle i : 0 \leq i \leq n :: A_i := A_i * B_i \rangle.$$

In many cases, programs can be structured in a variety of ways using combinations of assignments and statements, and assignments or statements can often be used alternatively. A very important difference, however, lies in the *execution* of the two constructs; let's clarify this with the above example, written as multiple assignment using $\|$ to separate the single components on one hand, and as a set of statements with $[\,]$ as separator between the statements. The execution of the *assignment version*

$$\| \langle i : 0 \leq i \leq n :: A_i := A_i * B_i \rangle$$

or in the explicit form (again with $n = 2$)

$$A_0 := A_0 * B_0 \parallel A_1 := A_1 * B_1 \parallel A_2 := A_2 * B_2$$

is achieved by evaluating *all* right-hand sides of the components *simultaneously*, and assigning the resulting values to the left-hand sides. As a consequence, data dependencies where variables appearing on the left side of an assignment are used in expressions on the right sight of another assignment can be fatal.

The *statement version*

$$[\,] \langle i : 0 \leq i \leq n :: A_i := A_i * B_i \rangle$$

or in the explicit form ($n = 2$)

$$A_0 := A_0 * B_0 \,[\,]\, A_1 := A_1 * B_1 \,[\,]\, A_2 := A_2 * B_2$$

is executed differently: here, *one* single statement is chosen *nondeterministically*, its right-hand side is evaluated and assigned to the left-hand side; data dependencies as mentioned above present no problem here.

The final section of a UNITY program specifies properties which hold as soon as a *fixed point* is reached[10]; in the above program such a property is that all paths must have been checked.

## Refinements

The program COMPLEMENTARY PATHS 1 has been derived in a quite straightforward manner from the basic features of the connection method; this is very nice with respect to the ease of program development, but suffers on the other hand from a lack of efficiency and maybe also of some elegance. In a second version shown in Figure 2.5 we introduce some improvements towards a more concise description.

The basic modification concerns the representation of the evaluation *result*; in the first version, this is done by explicitly assigning the values **proven**, **unprovable**, or **undefined**. This entails a number of relatively complex (pairs of) conditional assignments for the computation of the partial results and the final result. By employing the boolean values **true** and **false**, and interpreting them as 'proven' and 'unprovable' or 'undefined', respectively, the computation of $result_i$ becomes much simpler since the right-hand side of the corresponding assignment may consist of a boolean expression, which can be assigned directly to the variable. The determination of the overall result then simply consists of a conjunction of all partial results, mirroring the requirement that all paths have to be provable in order to prove a formula. Since the conjunction of the partial results always determines the overall result, it can be moved from the computation section to the invariant section. The only loss resulting from this modification is the fixed point property which states that all paths have been checked; since it is trivial anyway, this loss is not too severe.

The whole computation section of the program consists of one quantified assignment now, and the determination of the overall result is reduced from a statement containing two quantified comparisons for identity to a single boolean expression. Complexity considerations are much easier on the second version: the quantified assignment in the computation section requires as many instances as there are different paths through the matrix, and each assignment contains a computation of the *compl* predicate. The latter is quadratic to the length of the corresponding path, whereas the first is exponential with respect to the number of literals; then so is also the overall complexity for the

---

[10] sometimes also denoted *post conditions*

**Program** COMPLEMENTARY PATHS 2
**parameters**

| name | type | description |
|---|---|---|
| $F$ | logical formula | formula |
| $result$ | {true, false} | result of the evaluation |
| $\mathcal{PATHS}$ | set of paths | paths in the matrix |
| $PATH_i$ | set of literals | single path |
| $\mathcal{CLAUSES}$ | set of clauses | clauses of the matrix |
| $CLAUSE_j$ | set of literals | single clause |
| $\mathcal{LITERALS}$ | set of literals | literals of the matrix |
| $L_k$ | propositional variable | single literal |

**pre-conditions**

$$F \equiv \{CLAUSE_1, \ldots, CLAUSE_n\} \qquad \text{clause notation}$$

**local variables**

| $result_i$ | {true, false} | result for $path_i$ |
|---|---|---|

**invariants**

$$\forall\, i,j:\ PATH_i \in \mathcal{PATHS} \land PATH_j \in \mathcal{PATHS} \quad \text{all paths}$$
$$:: PATH_i \neq PATH_j \qquad \text{are different}$$

$PATH_i = \bigcup_{1 \leq j \leq\ |\mathcal{CLAUSES}|} L_j : L_j \in CLAUSE_j$  definition path: one literal from each clause

$compl(PATH_i) \equiv$  definition
$\exists\, k,l : L_k \in PATH_i, L_l \in PATH_i :: L_k = \neg L_l$  complementary path

$result = \bigwedge_{i \in \mathcal{PATHS}} result_i$  conjunction of partial results

**computation**

$\|\quad \langle i:\ PATH_i \in \mathcal{PATHS} ::$  for all paths
$\quad\ \langle result_i := compl(PATH_i) \rangle$  $path_i$ complementary ?

**end** COMPLEMENTARY PATHS 2

Figure 2.5: Program COMPLEMENTARY PATHS 2

evaluation of the formula; the generation of the paths is not regarded here.

As an important feature of a specification language, UNITY offers the possibility to reason about programs; it is possible, for example, to prove formally that a program reaches a certain state, or other interesting properties of a program. Since these constructs are not used very often in this book, they are introduced and explained at the time they are needed; for further details, see (Chandy and Misra, 1988).

## Processes

Processes represent the basic units in the description (of the operation) of complex computer systems; this concept is used as a specification tool to explain and understand the behavior of devices which we will refer to as *processors*, exemplified by digital computer systems. A process is defined based on the concepts of states and state variables (Horning and Randell, 1973; Chandy and Misra, 1988). A *state variable* is an elementary quantity which can assume certain well-defined values; the possible values for such a state variable are given in its *state space*. A *state* comprises a set of state variables together with *assignments* of values to all the variables; a *statement*[11] assigns values to a set of state variables simultaneously. A *computation* in a state space is a sequence of states from that space, and *operations* specify transitions from one given state to another state; one operation may consist of a number of statements. A *statement relation* in a state space is a mapping from states into statements, and is used to describe or generate a computation from an initial state to a terminal state[12]. If a statement relation is single-valued wherever it is defined, it is called *deterministic*, and actually represents a statement function; otherwise it is called *nondeterministic*.

### Definition 2.29 (Process)

A process is a triple $(\mathcal{S}, r, I)$, where $\mathcal{S}$ is a *state space*, $r$ is a *statement relation*, and $I$ is a subset of $\mathcal{S}$ specifying the *initial state* of the process[13]. A process is *deterministic* if its statement relation is deterministic, otherwise it is *nondeterministic*.                           ∎

To properly describe a processor based on the process concept we need to take into account two aspects. One is the *physical device* or machine (or an abstract model of it) which is capable of performing certain activities, assuming different physical statuses; the other is our *interpretation* of the activities performed by this device, consisting of a way to associate a mapping between the physical status of the pro-

---

[11] *action* in the terminology of (Horning and Randell, 1973)

[12] a UNITY program (see Section 2.2 specifies such a statement relation

[13] in UNITY, the state space is defined via the state variables used as listed in the parameter and local variable section, the statement relation through the statements of the computation section, and the initial state is given by the (imported) values of the parameters together with the values of the local variables as specified in the local variables section

cessor and the corresponding state of the process. In a typical digital computer the status of the processor is manifested via the contents of its registers described as binary words. The status of the processor can be interpreted as state of a process through software means (e.g. debugger), or by additional indicator devices like LEDs.

## Processes and Parallelism

The parallel execution of programs can be adequately modeled based on processes as defined above, and significant contributions to parallel processing have been made in this area (Horning and Randell, 1973; Ben-Ari, 1984; Hoare, 1985; Jorrand, 1986; INMOS, 1988b). These approaches describe parallel systems as networks of interacting processes. The composition of processes into networks relies on some primitive operators for synchronization, communication, and execution control, which form the basis for a formal treatment of such a network of processes, e.g. as an algebra of processes. As long as the structure of the system to be modeled can be described statically, this concept provides an adequate representation. To a limited degree, dynamically changing structures can also be modeled, e.g. by `fork` and `join` constructs, or by something like the turtle notation as used in LOGO (Pappert, 1980) and also proposed for Concurrent PROLOG (Shapiro, 1984).

# 2.3  Model Elimination

After the introduction to the connection method we will concentrate now on the actual evaluation mechanism used in a number of theorem provers, e.g. PTTP (Stickel, 1984), SETHEO (Letz et al., 1990a), PARTHEO (Schumann and Letz, 1990; Schumann, 1991). *Model elimination* (Loveland, 1968; Loveland, 1969; Loveland, 1979; Bayerl et al., 1989) can be viewed as a specialization of the connection method in the sense that it corresponds to one of the possible search attempts for complementary paths of a formula. An attempt is represented in the form of a *tree* or *tableau*, where different (sets of) paths are explored in different branches of the tree. The unsatisfiability[14] criterion is that each branch of the tree contains a pair of complementary literals (a connection): such a tree or tableau is called *closed*. An example for a model elimination tableau, using the same formula as in the previous section (Figures 2.1 – 2.3), is shown in Figure 2.6.

A specific condition for a *model elimination tableau* is that any node (except for the root and for the leaf nodes) has at least one complementary node among its direct successors. This restriction does not affect important properties like completeness, but helps to improve efficiency. The computation performed to obtain a proof for a certain formula is exactly the construction of a model elimination tableau; hence also its name: the closing of a branch corresponds to the elimination of a (class of) possible model(s) for the formula under investigation.

During the construction of such a tableau two essential steps have to be done repeatedly: *Extension*, which adds to an open branch of the tableau a clause containing a literal complementary to the one at the open end, thus adding another level to the tree. The branch containing the two complementary literals can be closed, and computation proceeds with still open branches. The second step is *reduction*, closing an open branch by a connection to one of the ancestors of the respective literal at the end of the open branch.

A necessary condition for both extension and reduction is that the two complementary literals to be connected are unifiable; in each step, the most general unifier for the pair of literals is computed, thus incrementally instantiating the variables involved. The whole proof process terminates successfully if all branches can be closed; if one or more

---

[14] a negative representation of the formula is used here

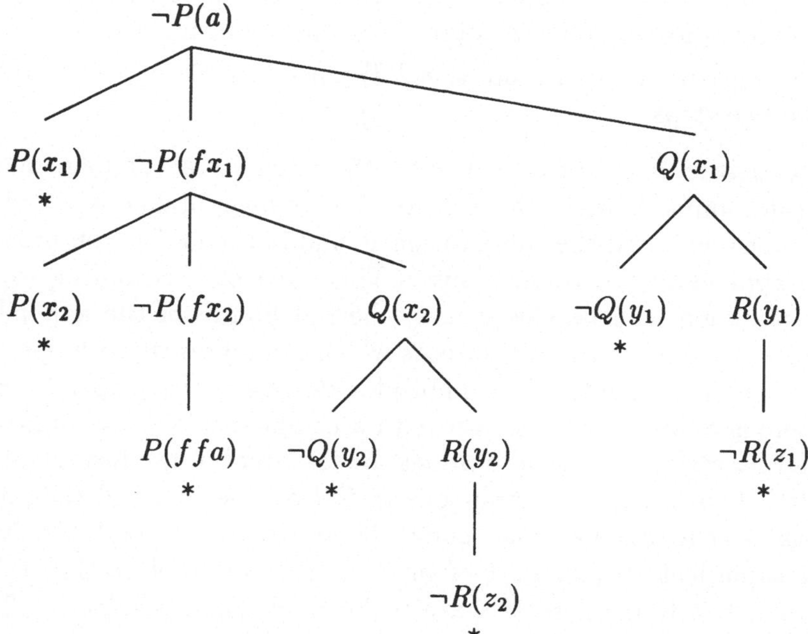

Figure 2.6: A model elimination tableau

branches cannot be closed, the proof (or refutation) attempt fails.

Model Elimination serves as basic evaluation mechanism for the inference mechanism of SETHEO and PARTHEO; from the implementation viewpoint, it also offers a sound basis for a procedural semantics (Letz et al., 1990a).

In the following, the description of an abstract inference machine based on the model elimination paradigm is given. It is specified using UNITY with possible parallelizations in mind; thus, the major point of interest is the identification of important features of model elimination with respect to a parallel evaluation, and it is not necessarily aimed at an efficient implementation.

The goal here is to design a program that computes the proof of a logical formula $F$. This is done according to *model elimination* by constructing a tree $PT_F$ representing a proof of the formula $F$. An auxiliary tree $T$ is constructed by appending appropriate clauses $C_i \in \{C_1, \ldots, C_n\}$ (the set of clauses of the formula $F$) at the ends of the branches of the tree. The local growth process stops when a branch of the tree is closed, i.e. it contains a literal together with its complement; the whole tree is computed as soon as all branches are closed. In this situation a fixed point is reached where $T = PT_F$ : the tree under construction $T$ is a proof tree for the formula $F$.

The construction of a proof tree $PT_F$ for a formula $F$ resides again on the two steps

1. **Extension:** In this step, a clause with a head literal complementary to (and unifiable with) the end literal of an open branch is added to the open end, and the substitutions found in the unification process are applied to the whole clause. This open end, according to its construction, can be closed now; the tail literals of the appended clause, however, form new sprouts which can represent both open or closed branches. If there are multiple extensions for one open branch (there is more than one clause with a complementary and unifiable head literal), alternative solutions for the proof of the formula may exist. This situation corresponds to OR-parallelism, and the point of the tree is called a choice point. These alternative solutions share the same history (the part of the tree from the root to the choice point), but from there on have to be treated independently. The conceptually easiest, but of course very memory-consuming way is to create an extra copy of the tree for each alternative.

2. **Reduction:** The purpose of the reduction step is to close as many open branches as possible in order to avoid unnecessary growth of the tree. The criterion for the closing of a branch is its complementarity, and in the extension phase branches are prolonged in a way that they can be closed by appending clauses with head literals complementary to the end literals. This complementarity with the end literal of a branch, of course, need not hold for the tail literals of the appended clause as well. The branches with the tail literals can be closed only if there is a complementary and unifiable literal in the same branch before the previous end point, and the corresponding substitutions have to be applied. As in the extension step, more than one possibilities may exist to close an open branch, introducing choice points and alternative solutions.

If all branches of the tree could be closed successfully, a proof for the formula has been found and is represented by the constructed tree. If there are open branches left which cannot be closed, this proof approach was not successful; alternative solutions, however, may

**Program** PROVER

**parameters**

| name | type | initially | description |
|------|------|-----------|-------------|
| $F$ | set of clauses | | formula to be proven |
| $\mathcal{PT}$ | set of trees | | proof trees |

**pre-conditions**

| | |
|--|--|
| $F \neq \{\}$ | formula is not empty |
| $\mathcal{PT} = \{\}$ | no proof for $F$ |

**local variables**

| | | | |
|--|--|--|--|
| $T_i$ | tree | $T_0 = \{\text{goal clause}\}$ | tree under evaluation |
| $stat(T_i)$ | $\{\text{closed}, \text{open}\}$ | **open** | status of a tree |
| $\mathcal{T_i}$ | set of trees | $\{\}$ | trees derived from $T_i$ |
| $\mathcal{T}$ | set of trees | $\{\}$ | trees under evaluation |

**computation**

$$\parallel \langle \quad i : T_i \in \mathcal{T} :: \qquad \text{for all trees}$$

$\quad\quad \parallel \boxed{\text{EXTENSION } (F, T_i, \mathcal{T_i})}$     extension step

$\quad\quad \parallel \boxed{\text{REDUCTION } (F, T_i, \mathcal{T_i})}$     reduction step

$\quad\quad \parallel \mathcal{PT} := \mathcal{PT} \cup \mathcal{T_i} \text{ if } stat(T_i) = \textbf{closed}$     add closed tree to proof trees

$$\rangle$$

**post-conditions**

| | |
|--|--|
| $F \neq \{\}$ | formula is not empty |

**end** PROVER

Figure 2.7: Program PROVER

exist. The formula cannot be proven only if all the trees representing alternative solutions have remaining open branches. The construction of a tree may also proceed infinitely and never tell us if the formula cannot be proven; this is due to the undecidability of predicate logic.

The program PROVER specifies the bare computational body for the evaluation of a formula: on a number of trees *extension* and *reduction* steps are performed. This is expressed by the two *statements* $\boxed{\text{EXTENSION } (F, T_i, \mathcal{T_i})}$ and $\boxed{\text{REDUCTION } (F, T_i, \mathcal{T_i})}$; the box around the statements means that they are described as separate programs later. According to the semantics of the specification language (Chandy and Misra, 1988), these two steps are executed independently and in a fair way in the sense that for an execution of the program which goes on forever, each statement of the program is selected infinitely often. In the sequential case, the selection is done nondeterministically. The quantification of statements denoted by '$\langle\ldots\rangle$' is an abbreviation, indicating that for each instance of the quantified variable a statement with the corresponding instantiations is created. Thus, for each tree under construction, extension and reduction steps are performed infinitely often in a nondeterministic way, or in parallel. Extensions and reductions in fact can be done independently, although in this case superfluous extension steps can occur; we will discuss later how to avoid them.

Let's have a closer look at the extension step now (see Figure 2.8): it takes one tree, and extends each branch by appending appropriate clauses from the formula under evaluation; in the case that there is more than one clause suitable for the extension of a branch, a copy of the tree is made for each clause. The extension of each of these trees is based on the extension of their single branches with one clause; these steps are described in $\boxed{\text{EXTREE } (NT_{ijk}, NB_{ij}, C_k, \sigma_{ijk})}$ .

The criterion for the extension of a branch with a clause is that the head literal of the clause is *complementary* to the current end of the branch; this is specified by an invariant saying that two literals are complementary if one is identical to the negation of the other, and they are unifiable. The computation part of EXTENSION expresses that for each candidate clause a copy of the original tree is made, and the corresponding clause is appended to the appropriate branch. The old branch $B_{ij}$ is stored temporarily in $NB_{ij}$ for further use in EXTREE, and $B_{ij}$ is removed from the tree to avoid multiple execution of the candidate clause collection. This method, of course, is extremely expensive in terms of space because a lot of copies of the tree are made, but it supports parallel evaluation by avoiding shared data structures.

The actual prolongation of the branches and the creation of new branches is specified by EXTREE. It takes a specific branch of a tree together with a clause and appends all literals of the clause at the end of the branch, creating new sprouts for each literal. In order to guarantee consistency of the substitutions for variables, unification is performed on the end literal of the branch and the head of the clause, resulting in a substitution which is applied to the whole clause as well as the tree under extension. The instantiations of the single literals then are appended to the end of the branch.

The goal of the reduction part is to detect all closed branches in a

## Program EXTENSION

**parameters**

| *name* | *type* | *initially* | *description* |
|---|---|---|---|
| $F$ | set of clauses | | formula to be proven |
| $T_i$ | tree | | tree to be extended |
| $\mathcal{T}_i$ | set of trees | | new trees from the extension of tree $T_i$ |

**pre-conditions**

| | |
|---|---|
| $F \neq \{\}$ | formula is not empty |
| $T \neq \{\}$ | tree is not empty |

**local variables**

| | | | |
|---|---|---|---|
| $NT_{ijk}$ | tree, set of branches | $\{\}$ | new tree, extension of branch $B_{ij}$ with clause $C_k$ |
| $NB_{ij}$ | sequence of literals | **empty** | temporary copy of $B_{ij}$ |
| $NB_{ijkl}$ | sequence of literals | **empty** | new branches) at the end of $B_{ij}$ for literals $L_{kl}$ |
| $C_{ij}$ | set of clauses | $\{\}$ | cand. clauses for extension of $B_{ij}$ |
| $\sigma_{ijk}$ | variable bindings | **nil** | substitutions for the extension of branch $B_{ij}$ with clause $C_k$ |

**computation**

| | |
|---|---|
| $[\!]\langle\ \ i,j : B_{ij} \in T_i \wedge B_{ij} ::$ | for all open branches |
| $\quad [\!]\qquad C_{ij} \quad := \{(C_k, \sigma_{ijk}) : \sigma_{ijk}(head(C_k))$ | collect cand. clauses |
| $\qquad\qquad\qquad\qquad = \sigma_{ijk}(end(B_{ij}))\}$ | and substitutions |
| $\quad\ \| \quad NB_{ij} \ := B_{ij}$ | copy branch |
| $\quad\ \| \quad B_{ij} \quad := \textbf{empty}$ | flush branch |
| $\quad\ \| \quad NT_{ij} \ := T_i \setminus B_{ij}$ | copy tree for extension |
| $\quad [\!]\quad \langle k : (C_k, \sigma_{ijk}) \in C_{ij} ::$ | for all candidate clauses |
| $\qquad [\!]\quad NT_{ijk} := \sigma_{ijk}(NT_{ij})$ | instantiate new trees |
| $\qquad [\!]\quad \boxed{\text{EXTREE } (NT_{ijk}, NB_{ij}, C_k)}$ | extend tree $NT_{ijk}$ at branch $NB_{ij}$ with $C_k$ |
| $\qquad [\!]\quad \mathcal{NT}_{ij} := \mathcal{NT}_{ij} \cup \{NT_{ijk}\}$ | add tree to growing trees |
| $\qquad\rangle$ | |
| $\quad [\!]\qquad \mathcal{T}_i \quad := \mathcal{T}_i \cup \{\mathcal{NT}_{ij}\}$ | collect new trees derived from $T_i$ |
| $\ \rangle$ | |

**end EXTENSION**

Figure 2.8: Program EXTENSION

**Program EXTREE**
**parameters**

| *name* | *type* | *initially* | *description* |
|---|---|---|---|
| $NT_{ijk}$ | tree | | tree to be extended |
| $B_{ij}$ | branch | | branch to be extended |
| $C_k$ | clause | | clause for extension |
| $\sigma_{ijk}$ | variable bindings | | substitution for the extension |

**pre-conditions**

| | |
|---|---|
| $NT_{ijk} \neq \{\}$ | tree not empty |
| $B_{ij} \neq \mathbf{empty}$ | branch not empty |
| $\sigma_{ijk}(head(C_k)) = \sigma_{ijk}(end(B_j\Gamma))$ | clause head and branch end are complementary |

**local variables**

| | | | |
|---|---|---|---|
| $NB_{ijkl}$ | sequence of literals | **empty** | new branches ('sprouts') at the end of $B_{ij}$ for literals $L_{kl}$ |

**computation**

$$\big[\big] \; \langle \quad l : L_{kl} \in C_k ::$$

for all literals of the clause

$$NB_{ijkl} := \sigma_{ijk}(B_{ij} \oplus L_{kl})$$

create new sprout

$$NT_{ijk} := NT_{ijk} \cup NB_{ijkl}$$

add new sprout to new tree

$$\rangle$$

**end EXTREE**

Figure 2.9: Program EXTREE

tree, and a fixed point is reached as soon as all closed branches are detected. For this purpose, each branch and each tree is assigned a status (**closed** or **open**). The status of a branch is said to be **closed** if there is a complementary pair of literals in the branch; a tree is **closed** if all its branches are closed. The following program is guaranteed to reach a fixed point since the number of branches to be checked is finite. In order to increase efficiency, the reduction can also be performed together with the extension of a branch of the tree, thus avoiding unnecessary extension steps, and limiting the check for complementary literals during reduction to a comparison of the end of the branch with the rest. For the sake of clarity, however, we first discuss the important features of the reduction step separately.

It is important to note that the computation performed during the

**Program** REDUCTION
**parameters**

| *name* | *type* | *initially* | *description* |
|---|---|---|---|
| $T_i$ | tree | | tree to be checked |
| $stat(T_i)$ | {closed, open} | | status of the tree |

**local variables**

| $B_{ij}$ | sequence of literals | **empty** | branches |
|---|---|---|---|
| $stat(B_{ij})$ | $\in$ {closed, open} | **open** | status of a branch |

**invariants**

stable($stat(T_i) = $ **closed**)                  a closed tree remains closed

stable($stat(B_{ij}) = $ **closed**)              a closed branch remains closed

**computation**

$\| \; \langle \quad j : B_{ij} \in T_i ::$            for all branches

$[\!]\quad stat(B_{ij}) := $ **closed** if $\exists\, L_m \in B_{ij} ::$    close

                $compl(L_m, L_{kl})$         branch

$[\!]\quad stat(T_i) \;\; := $ **closed** if $\forall\, B_m \in NT_{ijk} ::$    close

                $stat(B_m) = $ **closed**       tree

$[\!]\quad \mathcal{PT} \qquad\quad := \mathcal{PT} \cup T_i$         collect

                if $stat(T_i) = $ **closed**     proof trees

$\rangle$

**end** REDUCTION

Figure 2.10: Program REDUCTION

reduction step does not affect the trees and branches under evaluation: it only detects *properties* of these objects. For the detection of these properties new parameters and variables are introduced, the *status* of a tree or a branch, and in the reduction step only the values of these status indicators are changed. This allows us to place the statements of the reduction program to spots in the other programs where their execution increases efficiency.

After discussing the important properties of the essential parts of the model elimination proof mechanism separately, they will be included now into a single program. This will be done in two steps: first, the construction of the trees based on the extension steps is merged into one program, and then the necessary statements to determine important properties of the trees under construction are added. One of the most important properties here is if a branch is closed, detected by the reduction part; based on this, it is checked if a whole tree is

## Program EXTENSION-2

**parameters**

| name | type | initially | description |
|---|---|---|---|
| $F$ | set of clauses | | formula to be proven |
| $T_i$ | tree | | tree to be extended |
| $\mathcal{T}_i$ | set of trees | | trees derived from the tree $T_i$ |

**pre-conditions**

| | |
|---|---|
| $F \neq \{\}$ | formula is not empty |
| $T_i \neq \{\}$ | tree is not empty |

**local variables**

| name | type | initially | description |
|---|---|---|---|
| $\mathcal{NT}_{ij}$ | set of trees | $\{\}$ | trees created for alternative extensions of branch $B_{ij}$ |
| $NT_{ijk}$ | tree $\in \mathcal{NT}_{ij}$ | $\{\}$ | new tree created by extension of branch $B_{ij}$ with clause $C_k$ |
| $NB_{ij}$ | sequence of literals | **empty** | temporary copy of $B_{ij}$ |
| $NB_{ijkl}$ | sequence of literals | **empty** | new branches ('sprouts') at the end of $B_{ij}$ for literals $L_{kl}$ |
| $C_{ij}$ | set of clauses | $\{\}$ | cand. clauses for extension of $B_{ij}$ |
| $\sigma_{ijk}$ | variable bindings | nil | substitutions for the unification of the head of clause $C_k$ with the end of branch $B_{ij}$ |

**invariants**

| | |
|---|---|
| $\sigma_{ijk}(head(C_k)) = \sigma_{ijk}(end(B_{ij}))$ | head of $C_k$ and end of $B_{ij}$ are unifiable with substitution $\sigma_{ijk}$ |

Figure 2.11: a) Program EXTENSION-2 (Declaration Part)

closed, and if there are one or more proofs for the formula.

In order to achieve a unifying characterization of the extension part, we identify the essential activities performed in the previous code fragments EXTENSION and EXTREE. One is to identify for each (open) branch the set of candidates for extension among the clauses of the formula; another is to create a new tree for each of these candidate clauses, and a third is to append the literals of these clauses at the

**computation**

$[\!] \langle \quad j : B_{ij} \in T_i ::$  for all branches of $T_i$

$\quad [\!] \qquad\qquad C_{ij} \quad := \{(C_k, \sigma_{ijk}) :$  collect candidate

$\qquad\qquad\qquad\qquad \sigma_{ijk}(head(C_k))$  clauses and

$\qquad\qquad\qquad\qquad = \sigma_{ijk}(end(B_{ij}))\}$  substitutions

$\qquad\qquad\qquad\qquad \text{if } C_{ij} = \{\}$  (only

$\qquad\qquad\qquad\qquad \wedge B_{ij} \neq \textbf{empty}$  once)

$\quad \| \qquad NB_{ij} \quad := B_{ij}$  copy branch

$\quad \| \qquad B_{ij} \qquad := \textbf{empty}$  flush branch

$\quad \| \qquad NT_{ij} \quad := T_i \setminus B_{ij}$  copy tree for extension

$\quad \| \qquad T_i \qquad := \{\}$  flush old tree

$[\!] \langle \quad k : (C_k, \sigma_{ijk}) \in C_{ij} ::$  for all candidate clauses

$\quad [\!] \qquad\qquad NT_{ijk} \quad := \sigma_{ijk}(NT_{ij})$  instantiate

$\qquad\qquad\qquad\qquad \text{if } NT_{ijk} = \{\}$  new trees

$\quad [\!] \langle \quad l : L_{kl} \in C_k ::$  for all literals

$\quad\quad [\!] \qquad NB_{ijkl} := \sigma_{ijk}(NB_{ij} \oplus L_{kl})$  create a new sprout

$\qquad\qquad\qquad\qquad \text{if } L_{kl} \neq \textbf{nil}$  for each literal

$\quad\quad \| \; C_k \qquad := C_k \setminus L_{kl}$  remove used-up literal

$\quad\quad [\!] \qquad NT_{ijk} \quad := NT_{ijk} \cup NB_{ijkl}$  add sprout to new tree

$\quad\quad \| \; NB_{ijkl} := \textbf{empty}$  flush sprouts

$\quad\quad [\!] \qquad C_k \qquad := \{\} \text{ if } \forall L_{kl} \in C_k ::$  flush clause candidate

$\qquad\qquad\qquad\qquad L_{kl} = \textbf{nil}$  if all literals treated

$\quad\quad \rangle$

$\quad [\!] \qquad\qquad \mathcal{NT}_{ij} \quad := \mathcal{NT}_{ij} \cup \{NT_{ijk}\}$  add new tree

$\qquad\qquad\qquad\qquad \text{if } C_k = \{\}$  to growing trees

$\quad \| \qquad NT_{ijk} \quad := \{\}$  flush new tree

$\quad \| \qquad C_{ij} \qquad := \{\} \text{ if } \forall C_k \in C_{ij} ::$  flush set of

$\qquad\qquad\qquad\qquad C_k = \{\}$  candidate clauses

$\quad \rangle$

$\quad [\!] \qquad\qquad T_i \qquad := T_i \cup \{\mathcal{NT}_{ij}\}$  collect new trees

$\qquad\qquad\qquad\qquad \text{if } C_{ij} = \{\}$  derived from $T_i$

$\quad \| \qquad \mathcal{NT}_{ij} \quad := \{\}$  flush set of new trees

$\rangle$

**post-conditions**

$\forall T_i \in \mathcal{PT} :: stat(T_i) = \textbf{closed}$  all proof trees are closed

$\forall B_{ij} \in T_i, T_i \in \mathcal{PT} :: stat(B_{ij}) = \textbf{closed}$  all branches of a proof

  tree are closed

**end EXTENSION-2**

Figure 2.11: b) Program EXTENSION-2 (Computation Part)

appropriate spots, thus creating new branches ('sprouts'). These activities are described in the program EXTENSION-2; in addition, some 'housekeeping' has to be done to avoid multiple execution of statements (e.g. removing or marking the literals which have already been used for the extension of a branch).

The extension part being straightened up into a single program, the whole prover is derived now from the extension part. This is done by including the statements necessary to do the reduction steps on the trees under evaluation. These statements, as defined in the RE-DUCTION program, assign the appropriate status values to the single branches, or the whole tree. Their inclusion into the whole program – called PROVER-2 – can be done in two possible ways: One is to take the whole computation part of the REDUCTION program, and add it to the program EXTENSION-2; this variant will be described as PROVER-2a. The relevant parts are marked by an asterisk '*'; the first one closes a tree if all its branches are complementary; the second adds a tree to the set of proof trees if it is closed, and the other two are needed for housekeeping. In this case, these parts of the program can to a certain degree be treated independently, since they only manipulate data items (the status indicators) not used in the rest of the program; the trees and branches which have to be checked are not modified, but they must be accessible (either as shared data structures, or as copies). On the other hand, redundant extension steps may be performed since it is not guaranteed when a reduction step will be done. The other variant (described as PROVER-2b) is to integrate the relevant parts of the reduction step into the extension step at the appropriate place, making sure that a branch (or a tree) is checked for closedness immediately after creation, and thus no redundant extension takes place. The first assignment here closes a branch if the branch contains two complementary literals, and is again marked by '*'. The second marked statement has been moved into the statement which adds a new branch to a new tree to prevent unnecessary further extensions of this tree.

The second variant, PROVER-2b, performs reductions directly after the creation of new branches and trees, thus avoiding unnecessary extension steps. In the following, only the computation part is listed since there is no difference in the rest of the program to PROVER-2a.

The representation of a tree as a set of branches, with each branch given in its full length from the root to the current end point, obvi-

## Program PROVER-2a

**parameters**

| name | type | initially | description |
|---|---|---|---|
| $F$ | set of clauses | | formula to be proven |
| $\mathcal{PT}$ | set of trees | | proof trees |

**pre-conditions**

| | | | |
|---|---|---|---|
| $F \neq \{\}$ | | | formula is not empty |
| $\mathcal{PT} = \{\}$ | | | no proof for $F$ |

**local variables**

| | | | |
|---|---|---|---|
| $T$ | set of trees | $\{$goal clause$\}$ | 'growing' trees |
| $T_i$ | set of trees | $\{\}$ | trees derived from the tree $T_i$ |
| $\mathcal{NT}_{ij}$ | set of trees | $\{\}$ | trees created for alternative extensions of branch $B_{ij}$ |
| $NT_{ijk}$ | tree $in \mathcal{NT}_{ij}$ | $\{\}$ | new tree created by extension of branch $B_{ij}$ with clause $C_k$ |
| $stat(NT_{ijk})$ | $\{$closed, open$\}$ | **open** | status of a tree |
| $NB_{ij}$ | sequence of literals | **empty** | temporary copy of $B_{ij}$ |
| $NB_{ijkl}$ | sequence of literals | **empty** | new branches ('sprouts') at the end of $B_{ij}$ for literals $L_{kl}$ |
| $stat(NB_{ijkl})$ | $\{$closed, open$\}$ | **open** | status of a branch |
| $\mathcal{C}_{ij}$ | set of clauses | $\{\}$ | cand. clauses for extension of $B_{ij}$ |
| $\sigma_{ijk}$ | variable bindings | $\{\}$ | substitutions for the unification of the head of clause $C_k$ with the end of branch $B_{ij}$ |

**invariants**

| | | | |
|---|---|---|---|
| $\sigma_{ijk}(head(C_k)) = \sigma_{ijk}(end(B_{ij}))$ | | | head of $C_k$ and end of $B_{ij}$ are unifiable with substitution $\sigma_{ijk}$ |

Figure 2.12: a) Program PROVER-2a (Declaration Part)

ously is very expensive in terms of memory usage; on the other hand, however, it facilitates the evaluation in parallel since each branch can be treated independent of the rest of the tree. The same holds for the choice to represent alternative solutions of the formula through copies of the current tree.

**Program** PROVER-2a
**computation**

| | | |
|---|---|---|
| $[\!] \langle \; i,j : T_i \in \mathcal{T}, \; B_{ij} \in T_i \wedge B_{ij} \neq$ closed :: | | all trees and open branches |
| $[\!] \qquad C_{ij} \quad := \{(C_k, \sigma_{ijk}) : \sigma_{ijk}(head(C_k))$ | | collect cand. clauses |
| $\qquad\qquad\qquad = \sigma_{ijk}(end(B_{ij}))\}$ | | and substitutions |
| $\qquad\qquad$ if $C_{ij} = \{\} \wedge B_{ij} \neq$ empty | | (only once) |
| $\| \qquad NB_{ij} \;\; := B_{ij}$ | | copy branch |
| $\| \qquad B_{ij} \quad :=$ empty | | flush branch |
| $\| \qquad NT_{ij} \;\; := T_i \setminus B_{ij}$ | | copy tree for extension |
| $\| \qquad T_i \qquad := \{\}$ | | flush old tree |
| $[\!] \langle \; k : (C_k, \sigma_{ijk}) \in C_{ij} ::$ | | for all cand. clauses |
| $[\!] \qquad NT_{ijk} \;\; := \sigma_{ijk}(NT_{ij})$ if $NT_{ijk} = \{\}$ | | instantiate new trees |
| $[\!] \langle \; l : L_{kl} \in C_k ::$ | | for all literals |
| $[\!] \qquad NB_{ijkl} := \sigma_{ijk}(NB_{ij} \oplus L_{kl})$ if $L_{kl} \neq$ nil | | create new sprout |
| $\| C_k \qquad := C_k \setminus L_{kl}$ | | remove used literal |
| $[\!] \qquad NT_{ijk} \;\; := NT_{ijk} \cup NB_{ijkl}$ | | add sprout to new tree |
| $\| NB_{ijkl} :=$ empty | | flush sprouts |
| $[\!] \qquad C_k \qquad := \{\}$ if $\forall L_{kl} \in C_k :: L_{kl} =$ nil | | flush clause cand. |
| $\rangle$ | | |
| $[\!] \qquad stat(NT_{ijk}) :=$ closed if $\forall B_m \in NT_{ijk}$ | | close tree if all branches |
| $\qquad\qquad\qquad \exists L_n \in B_m :: compl(L_n, L_{kl})$ | | are complementary |
| $[\!] \qquad \mathcal{N}T_{ij} \;\; := \mathcal{N}T_{ij} \cup \{NT_{ijk}\}$ if $C_k = \{\}$ | | add tree to growing trees |
| $\| \qquad \mathcal{P}T \qquad := \mathcal{P}T \cup NT_{ijk}$ if $C_k = \{\}$ | | add tree to proof trees |
| $\qquad\qquad\qquad \wedge stat(NT_{ijk}) =$ closed | | if it is closed |
| $\| \qquad NT_{ijk} \;\; := \{\}$ | | flush new tree |
| $\| \qquad C_{ij} \quad := \{\}$ if $\forall C_k \in C_{ij} :: C_k = \{\}$ | | flush cand. clauses |
| $\rangle$ | | |
| $[\!] \qquad T_i \qquad := T_i \cup \{\mathcal{N}T_{ij}\}$ if $C_{ij} = \{\}$ | | collect new trees derived from $T_i$ |
| $\| \qquad \mathcal{N}T_{ij} \;\; := \{\}$ | | flush set of new trees |
| $[\!] \qquad \mathcal{T} \qquad := \mathcal{T} \cup \{T_i\}$ if $\forall; j \in T_i :: C_{ij} = \{\}$ | | collect all new trees |
| $\| \qquad T_i \qquad := \{\}$ | | flush set of trees |
| $\rangle$ | | |

**end** PROVER-2a

Figure 2.12: b) Program PROVER-2a (Computation Part)

**Program PROVER-2b**

**computation**

| | |
|---|---|
| $[\![ \langle\ i,j :\ T_i \in \mathcal{T},\ B_{ij} \in T_i \wedge B_{ij} \neq$ closed :: | all trees and open branches |
| $\qquad [\!] \qquad \mathcal{C}_{ij} \quad := \{(C_k, \sigma_{ijk}) : \sigma_{ijk}(head(C_k))$ | collect cand. clauses |
| $\qquad\qquad\qquad = \sigma_{ijk}(end(B_{ij}))\}$ | and substitutions |
| $\qquad\qquad\qquad$ if $\mathcal{C}_{ij} = \{\} \wedge B_{ij} \neq$ empty | (only once) |
| $\qquad \| \qquad NB_{ij} \ := B_{ij}$ | copy branch |
| $\qquad \| \qquad B_{ij} \quad :=$ empty | flush branch |
| $\qquad \| \qquad NT_{ij} \ := T_i \setminus B_{ij}$ | copy tree for extension |
| $\qquad \| \qquad T_i \quad := \{\}$ | flush old tree |
| $[\!] \ \langle\ k : (C_k, \sigma_{ijk}) \in \mathcal{C}_{ij} ::$ | for all cand. clauses |
| $\qquad\qquad NT_{ijk} \ := \sigma_{ijk}(NT_{ij})$ if $NT_{ijk} = \{\}$ | instantiate new trees |
| $\qquad \langle\ l : L_{kl} \in C_k ::$ | for all literals |
| $\qquad [\!] \ NB_{ijkl} := \sigma_{ijk}(NB_{ij} \oplus L_{kl})$ if $L_{kl} \neq$ nil | create new sprout |
| $\qquad \| C_k \qquad := C_k \setminus L_{kl}$ | remove used literal |
| * $\qquad \| \ stat(NB_{ijkl}) :=$ closed if | close branch if it has |
| $\qquad\qquad \exists\, L_m \in NB_{ij} :: compl(L_m, L_{kl})$ | complementary literals |
| $\qquad [\!] \ NT_{ijk} \ := NT_{ijk} \cup NB_{ijkl}$ | add sprout to new tree |
| $\qquad \| NB_{ijkl} :=$ empty | flush sprouts |
| * $\qquad \| \ stat(NT_{ijk}) :=$ closed if $\forall\, B_m \in NT_{ijk}$ | close tree if all branches |
| $\qquad\qquad \exists\, L_n \in B_m :: compl(L_n, L_{kl})$ | are complementary |
| $\qquad [\!] \ C_k \qquad := \{\}$ if $\forall\, L_{kl} \in C_k :: L_{kl} =$ nil | flush clause cand. |
| $\qquad \rangle$ | |
| $\qquad [\!] \qquad \mathcal{N}T_{ij} \ := \mathcal{N}T_{ij} \cup \{NT_{ijk}\}$ if $C_k = \{\}$ | add tree to growing trees |
| $\qquad \| \qquad \mathcal{P}T \quad := \mathcal{P}T \cup NT_{ijk}$ if $C_k = \{\}$ | add tree to proof trees |
| $\qquad\qquad\qquad \wedge stat(NT_{ijk}) =$ closed | if it is closed |
| $\qquad \| \qquad NT_{ijk} := \{\}$ | flush new tree |
| $\qquad \| \qquad \mathcal{C}_{ij} \quad := \{\}$ if $\forall\, C_k \in \mathcal{C}_{ij} :: C_k = \{\}$ | flush cand. clauses |
| $\qquad \rangle$ | |
| $\qquad [\!] \qquad T_i \quad := T_i \cup \{\mathcal{N}T_{ij}\}$ if $\mathcal{C}_{ij} = \{\}$ | collect new trees derived from $T_i$ |
| $\qquad \| \qquad \mathcal{N}T_{ij} := \{\}$ | flush set of new trees |
| $\qquad [\!] \qquad \mathcal{T} \quad := \mathcal{T} \cup \{T_i\}$ if $\forall; j \in T_i :: \mathcal{C}_{ij} = \{\}$ | collect all new trees |
| $\qquad \| \qquad T_i \quad := \{\}$ | flush set of trees |
| $\rangle$ | |

**end PROVER-2b**

Figure 2.13: Program PROVER-2b

The specification of the prover according to model elimination presented in the previous section has already been made in a way which easily allows for the extraction of parts to be treated in parallel. In the *prover* part, extension, reduction and the status check can be treated independently; the gain of performance, however, will not be very large since reduction and status check do not require very much computation, but a lot of communication if the trees are not kept in shared memory. A better choice is to combine extension, reduction, status check, and composition of the result into one program, and have it operate on the independent trees representing alternative solutions. This leads to a better balance between the underlying communication overhead and the computation tasks to be done in parallel.

The working packages are *tasks*, consisting of a partially constructed proof tree to be extended further. A new task is created for every new tree which comes into existence at a choice point. Since each of these trees has its own copy of the solution computed so far, and represents an alternative solution, its further extension can be pursued independently.

## 2.4   A Language: LOP

The initial impulse for the development of the language LOP (for LOgic Programming) was the necessity to provide an input language based on full first order logic for a parallel inference system (Bayerl et al., 1989). In the course of its definition and implementation, more features have been added to increase the ease of programming in general, and towards a specification and programming language for parallel systems in particular. However, this process is still in quite an early stage, and it might be somewhat premature to view LOP as a fully usable tool for the design and program development of parallel systems.

Its functionality and expressiveness comprise full first order logic, in contrast to PROLOG's restriction to Horn clauses, enhanced by a logically clean and concise integration of useful constructs known from conventional languages, like global variables and (destructive) assignment.

## The Syntax of LOP

The syntactical notation of the language LOP provides basically two variants, which can also be mixed within one program. One is similar to PROLOG notation, using the reverted conditional ':-' to separate the head of a clause from its tail. Here, a clause can be seen as a consequent (head), consisting of precisely one literal, which is implied by the antecedent (tail) consisting of a conjunction of literals. We use similar conventions as in PROLOG for constants, functions, predicates (starting with small letters), and variables (starting with a capital letter or underscore '_'). In contrast to PROLOG, literals can be negated as well, which is designated by a negation sign '¬', or a tilde ' ~ '.

The other variant, which allows a more concise description at the cost of reduced execution control, permits the use of multiple literals in the head of a clause. Formulae in mathematical logic notation can be specified with this variant in a straightforward way. In this case, the clauses of the program have to be "fanned out"[15] (Letz, ) to achieve a consistent evaluation for both notational variants. Due to this "fanning" and its accompanying decrease of control, a different symbol, namely '←' or '< −', is used in clauses with LOP-notation to separate head from tail. The typical form of clauses in PROLOG-like and LOP notation is shown in Figure 2.14.

$$
\begin{array}{llll}
head_1; \ldots; head_m & \leftarrow & tail_1, \ldots, tail_n. & \text{LOP rule} \\
head_1; \ldots; head_m & \leftarrow & . & \text{LOP fact} \\
& \leftarrow & tail_1, \ldots, tail_n. & \text{LOP query} \\
\\
head & \text{:-} & tail_1, \ldots, tail_n. & \text{PROLOG rule} \\
head. & & & \text{PROLOG fact} \\
& \text{?-} & tail_1, \ldots, tail_n. & \text{PROLOG query}
\end{array}
$$

Figure 2.14: PROLOG and LOP notation

A definition including both variants of the LOP syntax is given by the grammar[16] shown in figure 2.15. The syntax of LOP is designed such that both people with a background in PROLOG as well as in mathematical logic can easily use it. The PROLOG-like notation has

---

[15] fanning out a clause means to create an additional clause for each literal of the original clause, thus having the necessary entry points in the evaluation process; see Figure 2.17 for an example

[16] for the sake of readability, it is not really complete; it does not contain built-in predicates, for example

```
program  ::=    clauses
clauses  ::=    clause | clauses clause              one or more clauses
clause   ::=    lopclause '.'                         LOP notation
              | prlclause '.'                         PROLOG not.
lopclause ::=   '←' tail                              LOP query
              | lophead '←'                           LOP fact
              | lophead '←' tail                      LOP rule
prlclause ::=   '?-' tail                             PROLOG query
              | nliteral                              PROLOG fact
              | nliteral ':-' tail                    PROLOG rule
lophead  ::=    nliteral                              single literal
              | lophead ';' nliteral                  many literals
tail     ::=    nliteral                              single literal
              | tail ',' nliteral                     many literals
nliteral ::=    literal | '¬' literal                 positive or negated literal
literal  ::=    CONST                                 constant
              | CONST '(' parlist ')'                 function
              | GLOBVAR ':=' param                    assignment
              | GLOBVAR ':is' numterm                 forced eval.
              | VARIABLE 'is' numterm                 same value
              | binexpr                               relation
parlist  ::=    param                                 one parameter
              | parlist ',' param                     parameter list
param    ::=    CONST                                 constant
              | NUMBER                                number
              | VARIABLE                              variable
              | GLOBVAR                               global variable
              | STRING                                string
              | CONST '(' parlist ')'                 function
              | '[' list ']'                          list
list     ::=                                          empty list
              | parlist                               elements
              | param ' | ' param                     head | tail
numterm  ::=    NUMBER                                number
              | VARIABLE                              variable
              | GLOBVAR                               global variable
              | '(' numterm ')'                       parantheses
              | numterm '+' numterm                   addition
              | numterm '−' numterm                   subtraction
              | numterm '*' numterm                   multiplication
              | numterm '/' numterm                   division
              | '−' numterm                           negative
binexpr  ::=    numterm '>' numterm                   greater
              | numterm '<' numterm                   smaller
              | param '=' param                       equal
```

Figure 2.15: The Syntax of LOP

the advantage that it allows for a certain evaluation control through the textual arrangement of the clauses and literals; the LOP notation can be easily achieved from a mathematical logic formula.

## The Declarative Semantics of LOP

A meaning is given to a LOP formula $F$ by an *interpretation* $\mathcal{I}$, which maps the non-logical symbols of LOP into a certain object domain $D$; the logical symbols have their usual fixed meaning.

| LOP *construct* | $\Longrightarrow$ | *interpretation in the domain D* |
|---|---|---|
| constant | $\Longrightarrow$ | individual |
| variable | $\Longrightarrow$ | object |
| function symbol $f$ with arity $n$ | $\Longrightarrow$ | $n$-ary function $f' : D \times \ldots \times D \to D$ |
| term | $\Longrightarrow$ | object |
| predicate $P$ of arity $n$ | $\Longrightarrow$ | $n$-ary relation $P'$, or characteristic function $\mathcal{C}_P$ of $P'$ $\mathcal{C}_P : D \times \ldots \times D \to \{\text{true, false}\}$ |
| atomic formula $A(t_1, \ldots, f(t_n))$ | $\Longrightarrow$ | characteristic function $\mathcal{C}_A(t_1, \ldots, f(t_n))$, assigning a truth value to the formula |
| conjunction $\mathcal{I}(A \wedge B)$ | $\Longrightarrow$ | is **true** iff both $\mathcal{I}(A)$ and $\mathcal{I}(B)$ are **true** |
| disjunction $\mathcal{I}(A \vee B)$ | $\Longrightarrow$ | is **false** iff both $\mathcal{I}(A)$ and $\mathcal{I}(B)$ are **false** |
| universal quantification $I(\forall A(x))$ | $\Longrightarrow$ | is **true** iff for all objects $a$ in the domain $D$ $F(a)$ is **true** |
| existential quantification $\mathcal{I}(\exists A(x))$ | $\Longrightarrow$ | is **true** iff there is an object $a$ in the domain $D$ such that $F(a)$ is **true** |

Figure 2.16: An interpretation for LOP

For our purposes, a so-called *Herbrand interpretation* sufficiently characterizes the semantical properties of a LOP formula $F$. As object domain of $F$ we use its *Herbrand universe*, which consists of all terms constructed recursively from constants and function symbols occurring in $F$. The set of all atoms that can be built using predicate symbols of $F$ and the terms of its Herbrand universe is called the *Herbrand base* of the formula $F$. Any subset of the Herbrand base is a Herbrand interpretation, and the truth value of an atom is **true** if it is an element of a Herbrand interpretation.

A *model* $\mathcal{M}_F$ of a formula $F$ is an interpretation of $F$ such that $F$ has the value **true**; with these concepts – interpretation and model of $F$ – we can formulate an important distinction between a full first order LOP formula (or full LOP formula, for short), and a Horn clause formula (a 'pure' PROLOG program). For each Horn clause formula $H$ there is a *unique* minimal intersection of its Herbrand models $\mathcal{M}_H$ of $H$; this model can be regarded as minimal Herbrand model since the set of atoms (with the value **true**) is minimal, and in turn the Herbrand interpretation can be used to derive this minimal model. For full LOP formulae, there can be a number of different minimal intersections between its Herbrand interpretation and some models; a unique minimal intersection does not necessarily exist, which makes computation more complicated. The evaluation of full LOP programs can be made more efficient by a non-polynomial transformation ('fanning'), which derives an equivalent Horn clause program from each full first order logic formula. It is based on the observation that in a full LOP formula each literal of a clause can be used as 'entry point'[17]. A LOP clause can now easily be transformed into a set of Horn clauses, one for each literal. An example is given in Figure 2.17.

| *logical notation* | LOP *clause* | *fanned clauses* |
|---|---|---|
| $\neg p(x) \vee \neg p(y)$ | $\leftarrow p(X), p(Y).$ | $\neg p(X) : -p(Y)$ <br> $\neg p(Y) : -p(X)$ |
| $p(x) \vee p(y)$ | $p(X); p(Y) \leftarrow .$ | $p(X) : --\neg p(Y)$ <br> $p(Y) : --\neg p(X)$ |
| $p(x) \vee \neg(q \wedge r)$ | $p(X) \leftarrow q, r.$ | $p(X) : -q, r$ <br> $\neg q : --\neg p(X), r$ <br> $\neg r : --\neg p(X), q$ |

Figure 2.17: Fanning of a LOP formula

These considerations, together with the model elimination concept, also provide the basis for a procedural semantics of LOP described in the following.

---

[17] in a Horn clause program, only the single literal of the head can be used as 'entry point'

# The Procedural Semantics of LOP

The procedural interpretation of a LOP formula (Bayerl et al., 1989) is based on the connection method and model elimination as evaluation mechanism. The important steps during the construction of a proof for a formula according to model elimination are *extension* and *reduction* of the proof tree. An extension step corresponds exactly to a (binary) input resolution step, and hence can have the same interpretation as a procedure call; a reduction step basically can be seen as the termination of a procedure, making use of the parameters of its reduction partner.

Let's consider a typical situation in the proof process in more detail: An open node in the proof tree yearns to be closed; in clause terms, this is a subgoal to be solved, or a procedure to be evaluated in procedural terms. This node can be released by *extension* if a connection to a complementary literal in the head of a clause can be found, and a most general unifier exists. The connection to the clause head corresponds to the invocation of a procedure defined by this particular clause; the computation to be performed consists of the determination of the most general unifier according to the given parameters. Now the literals in the tail of this clause represent the new subgoals (invoked procedures) to be solved with the parameters initialized by the results of the previous unification.

In the *reduction* case an open node is not closed by looking for a new unification partner among the clause heads (procedure definitions), but among its predecessors in the path back to the root (the procedures already invoked on a higher level). The operation to close this node is again the determination of the most general unifier, using the actual parameters of the higher level procedure. The procedure under execution thus is terminated using the results previously computed by another instance of the same procedure.

This procedural semantics for LOP apparently only applies to its PROLOG-like form, where the head of a clause may consist of only one literal. The evaluation of full first order LOP programs is achieved by a transformation called *fanning* which maintains the essential logical properties of the formula (validity). Informally spoken, fanning creates for each literal of a clause a new clause with that literal as head; in the procedural interpretation, it serves as an 'entry point' for the clause. Fanning obviously increases the size of the formula considerably; its advantages, however, are fast access to connected subgoals, and the application of heuristics during compile time (e.g. reordering of clauses and subgoals within clauses). Fanning does not respect the notational order of the formula (i.e. the sequence of clauses and literals); thus it allows for less control from the user's side, potentially neglecting knowledge about the execution mechanism. The use of the LOP notation, which in the general case induces fanning, might be preferred by less experienced users, or in cases where the sequence is irrelevant.

## 2.5   Conclusions

The aim of this chapter was to introduce the foundations for an investigation of logic and parallelism. In the area of logic, the underlying concepts are well-defined with a good theoretical background, but the transposition to computer systems still suffers from lack of efficiency, experience, and sometimes 'cleanliness'. The usage of well-founded concepts for the design and programming of parallel systems is not very common, although some candidates exist or are under development (e.g. communicating sequential processes with Occam as implementation tool, Petri nets, FP2, UNITY). Although it was tempting to design and employ a logic-based language for the investigation of parallelism in logic, UNITY has been chosen since it is already in an advanced enough state of development to provide a sound basis for specification purposes. Its use turned out to be very helpful to achieve a better understanding for the problems investigated without having to deal with too many implementation details; it is not adequate, however, for the description of complex problems composed of smaller sub-problems, and puts some constraints on the way a problem is specified.

# Chapter 3

# State of the Art

This chapter reviews the major approaches in the area of parallelism and logic programming dealing with the parallel evaluation of logic programs; it covers parallel logic languages as well as parallel inference machines. The models proposed are discussed with respect to their *language* provided to the user, the *abstract machine* for the evaluation of programs written in that language, and *implementations* pursued or envisaged.

The description of parallel problems imposes great demands on the tools used; a *specification* should be clear, concise, and suited for formal reasoning, but still open to be used as basis for an efficient *implementation*. Logic possesses all the properties for a specification tool, but suffers considerably with respect to its efficient evaluation. Due to its declarative character, however, logic programming also provides an excellent basis for parallel evaluation, its semantics being (ideally) independent of the computational model and actual implementation. Its main representative PROLOG , however, deviates in at least two aspects from its logical basis; first, it is sensitive concerning the order of its evaluation, and second it contains some non-logical features (esp. the `assert` construct). The reasons for these 'blunders' are obvious requirements on a programming language: efficiency, and adequate handling of input and output operations. As a consequence, parallel implementations based on PROLOG face the problem to either maintain the operational semantics resulting in restricted exploitation of parallelism, or to renounce compatibility to the currently most renowned representative of logic programming languages.

## 3.1   Parallel Logic Systems

The goal of this section is to present a short overview of computer systems with *logic* as underlying programming paradigm, and aimed at the exploitation and description of *parallelism*. Some of these proposals are aimed at integrated systems in the form of *parallel inference machines*, featuring logic programming languages with all necessary development tools running on dedicated hardware.

### Languages

The main problems with the use of mathematical logic for *programming*, lack of efficient evaluation mechanisms and limited execution control, have been partially overcome by PROLOG, trading in restricted expressiveness and questionable semantics for higher efficiency, better execution control, and handling of I/O. The combination of logic and parallelism is inspired by two main motivations; one is an increase of performance via parallel evaluation of logic programs, the other an appropriate specification of parallel systems with logical means.

With respect to parallelism, three basic types of logic languages can be identified:

**"Pure" Logic Languages** are advocated as integrated tools for the specification, validation, and implementation of models for parallel systems. Approaches in this direction are usually based on pure (predicate or Horn clause) logic, with enhancements for the description of the structure of parallel systems, and their behavior (Goguen and Meseguer, 1987; Kurfeß and Schumann, 1989). These enhancements in principle require arguing about predicates, and thus higher order logic. These concepts have an impressive elegance and clarity, but still require considerable efforts to achieve practical usability. Especially their performance has to be increased a lot, which is intended by the adaptation of parallel evaluation, making use of various kinds of parallelism. A number of problems also arise with respect to denotational and operational semantics.

**"All-Solutions" Languages** are attempts to parallelize (sequential) PROLOG while maintaining its semantics as is. This task is severely impeded by PROLOG's evaluation mechanism, which is order-dependent and contains some constructs with side-effects. The reason for this is the transformational nature of the PROLOG evaluation model, receiving an input at the beginning, doing computations, and yielding an output value at the end. Input / output facilities are not easily integrated, but rather an ad hoc addition.

The main source of parallelism is the independent search for different solutions (OR-parallelism), sometimes accompanied by a restricted exploitation of AND-parallelism (de Groot, 1984). A problem is the management of the variable bindings for the different solutions, which are organized in a tree-like manner. Since usually the available resources are restricted, the full potential of parallelism may not be exploited, and provisions for sequential execution (backtracking) have to be included, which again increases the implementation complexity. This approach is the most popular one, and in addition to the projects reviewed in this section (GIGALIPS, ANDORRA, PEPSYS), many more have been proposed (Haridi and Ciepielewski, 1983; Ponder and Patt, 1984; Tick and Warren, 1984; Voda and Yu, 1984; Matsuda and Kokata, 1985; Crammond, 1986; Ali, 1987; Warren, 1987; Shen and Warren, 1987).

Most of these proposals actually use (a subset of) PROLOG as language, possibly enhanced by some explicit parallel constructs, and derive parallel computational models from the resolution calculus as well as abstract machines from Warren's abstract PROLOG machine (Warren, 1983).

**"Committed Choice" Languages**   (also called concurrent logic programming languages) are derived from a common ancestor, the *relational language* (Clark and Gregory, 1981). These languages have a syntax based on Horn clauses and thus similar to PROLOG, but feature a quite different operational model, and a programming style which is relatively close to procedural programming. The main representatives are PARLOG (Clark and Gregory, 1986), Concurrent PROLOG (Shapiro, 1986; Shapiro, 1988), and GHC (Guarded Horn Clauses) (Ueda, 1985); (Shapiro, 1989) contains an excellent overview of these languages.

Programs are sets of *guarded Horn clauses* of the form

$$H : -G_1, \ldots, G_n \mid B_1, \ldots, B_m.$$

The *commit* operator $\mid$ separates the *guard* from the *body* part of the clause; the function of the guard part is to select the clause to be invoked for the computation of a predicate. Once a clause has been selected, it must be used, and there is no way of undoing the selection, as with backtracking; this is why these languages are also called "single-solution" languages. The selection of a clause is done non-deterministically, and may lead to different results for different executions of one and the same program. Advocates of this model, however, point out the correspondence of committed choice non-determinism to input / output operations: once a message is sent, there is no way to "unsend" it. The responsibility for making choices is in the hands of the programmer, and not dependent on the language implementation.

The parallelism exploited here is mainly of the AND type, with some OR during the evaluation of the guard parts. Especially for synchronization purposes, extensive use is made of shared variables; to guarantee consistency, syntactical constructs like mode declarations (PARLOG) and read-only variables (Concurrent PROLOG), or semantical rules (GHC) are required. Since the permission of arbitrary goals in the guard part leads to a (possibly very complex) process hierarchy in the implementation, *flat* versions of these languages have been developed, where only predefined primitive operations may be performed in the guard parts. As a consequence, the execution model consists of a 'flat' pool of processes; these processes have a single environment and no hierarchical connections.

Most of these languages are still in the research or prototype status; the only commercially available parallel logic languages are CS-PROLOG (Futo and Kacsuk, 1989) and STRAND88 (Foster and Taylor, 1989).

## Computational Models

A *computational model* for a logic language describes the process of evaluating a program of that language; its main influence comes from the underlying logical calculus used. The most popular computational models have been derived from the resolution calculus (Robinson, 1965) used for PROLOG. Also of considerable interest is the Connection Method (Bibel, 1987), providing the basis for Model Elimination as used in SETHEO and PARTHEO, the Spanning Setters described in this book, and the related proposals of Ibanez (Ibañez, 1989) and Wang's Wivenhoe model (Wang, 1989).

Another common starting point for a computational model is to view the evaluation of a logic program as traversal of a search tree; of particular interest here is the AND/OR tree model (Conery, 1983; Gregory et al., 1989).

## Abstract Machines

An *abstract machine* provides an evaluation mechanism for the computational model of a logic language without considering the details of an actual implementation on a specific machine; frequently evaluation mechanisms are provided in the form of an interpreter with the same behavior as the abstract machine, or by a transformation (compilation) into abstract machine instructions. These instructions can be executed by an emulator (a program written in C, for example), via microcode, or via a compilation into the machine code of the underlying hardware. In some cases intermediate or restricted versions of languages are defined in order to obtain a less complicated abstract machine, and high-level programs are compiled, maybe in several steps, into lower-level programs for a particular abstract machine.

The most prominent abstract machine for the evaluation of logic programs is D.H.D. Warren's PLM (Programmed Logic Machine), better known as WAM (Warren Abstract Machine) (Warren, 1983). Although initially designed for sequential operation, it provides the basis for many parallel implementations; because of this impact, its main characteristics, extended by some concepts concerning parallelism, are summarized in Table 3.1.

| EXTENDED WARREN ABSTRACT MACHINE | | |
|---|---|---|
| | data structures | operations |
| **compu-tation** | PROLOG **term** (value + tag) | **unify** instructions<br>**procedural** instructions<br>**indexing** instructions |
| **repre-sen-tation** | **code area** (PROLOG program): sequence of PROLOG instructions<br>**stack** (local): contains environments and choice points<br>**heap** (global): contains structures and lists created by unification and procedure invocation<br>**trail:** contains references to variables bound during unification; unbound by backtracking) | **get** instructions<br>**put** instructions<br>(**indexing** instructions) |
| **com-muni-cation** | PROLOG **terms** (value + tag) | **send** instructions<br>**receive** instructions |

Table 3.1: Extended Warren Abstract Machine

## Hardware Architectures

Dedicated architectures and devices for fast execution of logic languages have been proposed both for sequential and parallel approaches, most of the sequential ones being also usable at least as multi-sequential execution mechanisms by means of interconnection and software. The parallel approaches range from dedicated building blocks for the execution of particular operations (e.g. unification, fast access to literals) like PAM (Robinson, 1986) or (Georgescu, 1986), over coprocessors for conventional workstations like PLM (Citrin et al., 1986; Van Roy, 1991), POPE (Beer and Giloi, 1987; Beer, 1989), ICM-4 / KCM (Watzlawik et al., 1987), to self-contained PROLOG machines like PSI (Taki et al., 1984). Many systems and building blocks designed for more general purposes (like symbolic programming) or applications with similar requirements (functional programming, dataflow, object-oriented) may provide suitable execution platforms for logic programs as well (Treleaven, 1989).

## Inference Machines

The term *parallel inference machines* is used for implementations of parallel logic languages on parallel computer systems in an integrated way; in the extreme case the architecture of the system is specifically designed for and dedicated to the execution of parallel logic programs. The most ambitious project has been pursued in the Japanese Fifth Generation Computer Systems framework, aiming at a high performance parallel system for information processing purposes (Parallel Inference Machine – PIM) (Goto et al., 1988). Also in Japan, a Kabu-Wake multiprocessor system consisting of 16 standard microprocessors (MC68010) has been built (Sohma et al., 1985).

Most of the other implementations of parallel logic systems are either based on commercially available multiprocessor systems (e.g. Concurrent PROLOG on a Intel iPSC Hypercube, PEPSYS on a Siemens MX500, some more on similar bus-based UNIX multiprocessors, PARTHEO on Transputers), or integrated within larger research projects (e.g. PARLOG and FLAGSHIP (Treleaven, 1989), FOOPSLOG within the Rewrite Rule Machine project (Goguen et al., 1989), the Parallel PROLOG Processor within Aquarius (Fagin and Despain, 1987)).

A number of other research projects contain parallel inference mechanisms as essential components, e.g. FAIM (Anderson et al., 1987), PRISM (Kasif et al., 1983), MANIP-2 (Li and Wah, 1985). In addition, parallel implementations of logic languages are pursued in smaller projects on various parallel systems.

On the following pages, important systems – languages, abstract machines, and implementations – are briefly described, and their important characteristics concerning language, abstract machine and implementation are presented in overview tables.

# ANDORRA

The goal of ANDORRA (Haridi and Brand, 1988) as parallel execution model for PROLOG is to integrate both PROLOG and committed choice languages, making *determinism* the basis for transparent exploitation of dependent AND-parallelism and OR-parallelism. In this model, determinate goals – which have at most one candidate clause that matches – are executed in AND-parallel as long as there are determinate goals left; OR-parallelism occurs if all remaining goals are non-determinate.

The Andorra execution model relies on two principal operations, AND-reduction and OR-extension, which are applied to the tree of nodes describing the computation. AND-reduction is used for determinate goals and has a further potential for parallelization in its elementary operations dealing with unification, the publication of variable bindings, the reduction of goals, and the reduction of candidate clauses.

The degree of OR-parallelism is essentially the same as in OR-parallel systems; for determinate logic programs, stream and independent AND-parallelism are equivalent to committed choice languages.

The language derived from this model, Andorra PROLOG, supports don't know- and don't care-nondeterminism, as well as control of OR-parallelism, synchronization on variables, and clause selection. The ability to express don't know-nondeterminism is used for the exchange of information between objects by time-stamped messages, which can be used to describe distributed discrete event simulation problems, for example.

It supports PROLOG with some limitation of the cut operator, and programs written in (flat) committed choice languages can be expressed in a straightforward way. A full incorporation of the cut operator, however, would severely restrict AND-parallelism.

In contrast to committed choice languages, the Andorra model does not require special language constructs to express and control parallelism. There is also no restriction to generate one solution only.

# ANDORRA

### *Language*

**dialect:** Andorra PROLOG

**scope:** Horn clauses

**calculus:** resolution

**computational model:** Andorra parallel execution model with AND-reduction and OR-extension as essential operations

**parallelism:** AND-parallelism, stream parallelism, OR-parallelism, parallelism on lower levels (selection of candidate clauses, publication of bindings, unification completion)

**control of parallelism:** implicit determinate and nondeterminate goals; *guards* to restrict selection of clauses, *commit* operator '|'(symmetric cut, pruning the branches both to the left and the right), *OR-cut* maintaining the standard *cut* semantics in OR-parallel execution, but relaxing it in AND-parallel mode, *delay* predicate for blocking the execution of a goal, and synchronization of variables

### *Abstract Machine*

abstract execution mechanism based on the construction of a tree, a *configuration* comprises the list of remaining goals, and a mode (AND, OR, failure); AND-reduction / OR-extension are applied to the nodes of the tree

### *Implementation*

the techniques developed in the AURORA (cf. 3.4) OR-parallel system are envisaged

Table 3.2: ANDORRA

# AQUARIUS / PPP

The AQUARIUS project is a research project headed by Alvin M. Despain and Yale N. Patt at the University of California, Berkeley; it is aimed at architectural investigations of high performance computation (Despain and Patt, 1985; Fagin and Despain, 1990). PROLOG being the user language of the system, a parallel execution model for PROLOG, PPP (Parallel PROLOG Processor) (Fagin and Despain, 1987) has been designed, based on previous research on a high performance sequential PROLOG architecture, the PLM (PROLOG Machine) (Srini et al., 1987). The sequential architecture has been implemented on one chip (VLSI-PLM), and a commercial version is available as Sun coprocessor (Xenologic X-1). The most recent development is the VLSI-BAM (Berkeley Abstract Machine), which in combination with the corresponding optimizing compiler is about 10 times faster than the VLSI-PLM. Retargeted to the SPARC, it outperforms Quintus PROLOG by a factor of five (Van Roy, 1991).

The PPP aims at improved performance for (standard) PROLOG by combining restricted AND-parallelism, OR-parallelism, and intelligent backtracking. It is implemented through compilation to an abstract PROLOG instruction set. These abstract instructions are executed on a multiprocessor simulator for an arbitrary number of processors, each processor being an extended version of the Berkeley PLM PROLOG processor (Citrin et al., 1986). Included in the project is also work on microarchitectures for fine-grain parallelism in the instruction stream. This does not only cover unification, but fine-grain parallelism for any deterministic goal (Singhal, 1990; Van Roy, 1991). Considerable effort has been put into a reduction of the overhead during execution of a program without enforcing explicit control by the programmer, aiming at high performance in absolute terms, not only in comparison with possibly inefficient sequential systems.

A detailed analysis of the performance of parallel PROLOG programs based on the PPP execution model is given in (Fagin and Despain, 1990). The main results are that restricted AND-parallelism showed significant speed up only on two large programs, and OR-parallelism was basically not useful if the semantics of PROLOG were preserved. An interesting phenomenon was the supermultiplicative behavior obtained when more than one technique (restricted AND-parallelism, OR-parallelism, intelligent backtracking) was applied.

<table>
<tr><td colspan="2" align="center"><h1>AQUARIUS / PPP</h1></td></tr>
</table>

*Language*

**dialect:** (standard) PROLOG
**scope:** Horn clauses
**calculus:** resolution
**computational model:** AND / OR-processes, intelligent backtracking
**parallelism:** restricted AND-parallelism identified at compile time, OR-parallelism, fine-grain parallelism
**control of parallelism:** mostly implicit, but supports annotations for AND- and OR-parallelism

*Abstract Machine*

**type:** extension of Warren's abstract machine
**topology:** AND / OR-tree of processes
**communication mechanism:** messages
**task management:** global process table
**parallelism:** independent processes
**control of parallelism:** hierarchy of processes, managed by a process kernel

*Implementation*

**implementation language:** C
**operating system:** UNIX
**host:**
**node:** modified PLM processor
**topology:** two: tier: full interconnection (crossbar), bus with consistent caches (for synchronization)
**communication:** shared memory
In (Fagin and Despain, 1987), only a simulator for the PPP is described, based on the Aquarius architecture; the results reported in (Fagin and Despain, 1990) were also obtained via simulations. VLSI and commercial implementations are available for the sequential versions.

Table 3.3: AQUARIUS / PPP

## AURORA

The AURORA system has been developed by three groups of researchers from the Mathematics and Computer Science Division of the Argonne National Laboratory, Department of Computer Science of Manchester University, and the Swedish Institute of Computer Science within a research collaboration known as "Gigalips Project"(Lusk et al., 1988). It is a prototype OR-parallel implementation of full PROLOG, based on Sicstus PROLOG, which in turn is a C implementation of Warren's abstract machine. The purposes of the system are to enhance the understanding of the requirements for parallel logic system, and to gain experience with the parallel evaluation of large applications. For easy use of existing applications, AURORA contains full PROLOG with its standard semantics as a true subset of the language.

A number of reasons have influenced the decision to focus exclusively on OR-parallelism; *generality* to maintain the power of the underlying logic language, here in particular the ability to generate all solutions to a goal; *simplicity* to avoid any explicit annotation or complicated analysis of parallelism; *closeness to* PROLOG to take advantage of existing implementation technology; *granularity* to generate sufficiently large-grained processes for current multi-processor systems, and finally a substantial amount of *applications* exhibiting a sufficient degree of OR-parallelism.

The representation of different bindings of the same variable during the execution of a program is the main problem with OR-parallelism. Each worker has a private binding array to record conditional bindings; the WAM stacks are extended to "cactus stacks", reflecting the shape of the search tree. The objects in the physical stacks of workers may be linked, forming a logical stack.

Preliminary performance results (Lusk et al., 1988) show a speedup of up to 14 (relative to the speed on one processor) for smaller programs with a high degree of OR-parallelism; larger programs taken from real applications also show a substantial speedup. On a single processor, AURORA is about 25% slower than its sequential relative Sicstus.

<table>
<tr><td colspan="1" align="center">

# Aurora

### *Language*

**dialect:** full PROLOG (standard semantics), plus additional constructs
**scope:** Horn clauses
**calculus:** resolution
**computational model:** SRI model; a group of workers cooperate to explore the search tree
**parallelism:** OR-parallelism
**control of parallelism:** mainly implicit, some special predicates (`:- sequential < procedure> <arity>`) to restrict parallelism

### *Abstract Machine*

**type:** extended WAM
**topology:** tree
**communication mechanism:** shared data (public part of the search tree, global binding array)
**task management:** schedulers release work (top node of a branch, breadth-first strategy), idle workers traverse the search tree for work; suspension if a *cut* or side-effect predicates cannot be executed
**memory management:** cactus stacks; binding arrays
**parallelism:** OR-parallelism
**control of parallelism:** tasks, choice points private and public parts of the search tree

### *Implementation*

**system:** Sequent Symmetry, Encore Multimax
**implementation language:** C
**operating system:** UNIX
**host:** SUN
**node:** Intel 80386 (Symmetry)
**topology:** bus
**communication:** shared virtual address space

</td></tr>
</table>

Table 3.4: Aurora

## Concurrent Prolog

Concurrent PROLOG has been designed by Ehud Shapiro at the Weizmann Institute in Rehovot, Israel (Shapiro, 1983; Shapiro, 1988). As a member of the family of concurrent logic programming languages, it preserves many features of the abstract logic programming model, while it also offers basic synchronization and control constructs for programming parallel systems. The operational model consists of a dynamic concurrent processes, exchanging information through shared variables, and synchronized through the instantiation of variables.

Its development is in close contact with both the work on GHC at ICOT (Shapiro and Takeuchi, 1983; Takeuchi and Furukawa, 1985; Ueda, 1985) and the activities on PARLOG at Imperial College, London (Foster and Taylor, 1987). The main feature differentiating Concurrent PROLOG from GHC and PARLOG is the treatment of variables and unification, and the expression of synchronization. PARLOG and GHC only allow the testing of variable bindings during the evaluation of the guard part; Concurrent PROLOG permits variables to be bound in this phase, which requires the maintenance of multiple environments and may lead to incorrect bindings if it is subsequently decided that this clause cannot be used.

Whereas the mechanism used in Concurrent PROLOG facilitates some sophisticated programming techniques, their implementation is afflicted with some problems (Foster and Taylor, 1987; Saraswat, 1986), e.g. the necessity to implement unification as atomic operation. For the expression of synchronization constraints (data flow), Concurrent PROLOG uses read-only variables, indicating that processes trying to unify a read-only occurrence of a variable are suspended until it is bound. Procedurally spoken, the constraints are given in the call to a procedure rather than in the procedure itself; this requires either a global program analysis or run-time checks for the use of matching or full unification.

Full Concurrent PROLOG requires a highly complex and computationally intensive treatment of multiple variable bindings; for this reasons, *Flat* Concurrent PROLOG has been introduced, restricting the predicates allowed in the guard part to a fixed finite set of system predicates.

# CONCURRENT PROLOG

### *Language*

**dialect:** committed choice language

**scope:** Horn clauses

**calculus:** reduction

**computational model:** guarded Horn clauses; from a set of alternative clauses, one is selected according to the conditions defined in the guard part; after the commitment to a clause, its body part is evaluated; only one solution is computed, and no backtracking is possible

**parallelism:** mainly AND-parallelism, OR-parallelism only in the evaluation of the guard part, stream parallelism

**control of parallelism:** mainly implicit, sequentialization through guards and read-only variables, synchronization via read-only variables

### *Abstract Machine*

**type:** derived from Flat Concurrent PROLOG, (only built-in system predicates in the guard part)

**topology:** flat pool of processes

**communication mechanism:** shared variables

**task management:** process structure according to the AND-/OR-tree of the reduction mechanism, suspension of processes if variables are not instantiated

**parallelism:** implicit

**control of parallelism:** implicit

### *Implementation*

**implementation language:** C, Occam

**operating system:** UNIX

**node:** standard microprocessors (80x86, 680x0, Transputer)

**topology:** hypercube, torus

**communication:** message passing

Table 3.5: CONCURRENT PROLOG

## Delta Prolog

Delta PROLOG (Moniz Pereira et al., 1988; Cunha et al., 1989) is
an extension of (sequential) PROLOG to include AND-parallelism and
inter-process communication. In contrast to other parallel logic languages based on AND-parallelism, Delta PROLOG subsumes PROLOG,
does not require a synchronization mechanism with different unification semantics, and does not enforce commitment.

On the language level, three additional goal types are introduced:

**split goals** of the form $S_1//S_2$, where $S_1$ and $S_2$ are arbitrary goal
expressions. Declaratively, $S_1//S_2$ is **true** iff $S_1$ and $S_2$ are jointly
**true**, and operationally they are solved by asynchronous parallel processes.

**event goals** of the form $X?E : C$ or $X!E : C$, where $X$ is a term
(*message*), $E$ a PROLOG atom (*event name*), and $C$ is a goal expression (*event condition*) which may not evaluate further goals. Event
goals can be unified if they have the same name, and their *communication modes* ? or ! are complementary[1].

**choice goals** of the form $A_1 :: A_2 :: \ldots :: A_n$, where :: is the *choice
operator*, and $A_i$ the *alternatives* of the goal. Alternatives in the
form of choice goals are used to express *external* or *global non-
determinism*, in contrast to *internal* or *local non-determinism*, which
is modeled by separate clauses potentially to be unified with a goal.

A declarative semantics for Delta PROLOG is derived from a Herbrand
interpretation enhanced by a concept of event histories ("traces") to
capture the additional goals introduced. An operational semantics
is based on an exploration of the derivation space, starting from the
top goal with a *root* process. It proceeds depth-first until a split goal
is met, where an additional process is spawned: Thus coordinated
execution of distinct threads of control, represented by processes[2],
provides the computation for each solution.

---

[1] (which does not necessarily specify a direction like *read* and *write*)

[2] separate interpreter instances of C-PROLOG in the current implementation

# DELTA PROLOG

*Language*

**dialect:** extension of PROLOG

**scope:** Horn clauses

**calculus:** resolution

**computational model:** construction of a derivation via a tree-structured network of coordinated processes

**parallelism:** AND-parallelism

**control of parallelism:**

- *split* goals for AND-parallelism,
- *event* goals for inter-process communication,
- *choice* goals for external non-determinism

*Abstract Machine*

**type:** extension of WAM

**topology:** tree

**communication mechanism:** term unification

**task management:** spawning of processes when split goals occur

**parallelism:** AND-parallelism

**control of parallelism:** coordinated processes

*Implementation*

**emulation** based on an asynchronous execution of multiple instances of an extended C-PROLOG

Table 3.6: DELTA PROLOG

## GHC – Guarded Horn Clauses

Based on experiences with Concurrent PROLOG, GHC (Ueda, 1985) represents the intermediate language of the Japanese Fifth Generation Computer Systems project (Uchida et al., 1988; Goto et al., 1988). GHC serves as a basis for general purpose and knowledge programming languages, and is executed on the (multi-) PSI and PIM hardware developed in the same project.

Similar to the other committed choice languages Concurrent PROLOG and PARLOG, GHC uses the concept of guarded Horn clauses with the commit operator to implicitly specify parallel programs. Unification of variables is used for synchronization and communication purposes. The correctness of bindings is guaranteed by semantical rules restricting unification in the head or guard part of a clause to non-modifying tests; a unification in the head or the guard that tries to instantiate a variable in the goal is suspended. PARLOG achieves the same using syntactical mode declarations, whereas Concurrent PROLOG allows full unification in the head and guard part, leading to better expressiveness connected with a more complicated implementation.

The implementation of full GHC also imposes considerable run-time constraints on the data structures and algorithms. As remedies, two subsets of GHC were identified: *flat* GHC and *safe GHC*. The flat version restricts the use of predicates in the guard part, and the safe version does not allow executions in which a body unification suspends. A flat GHC program is always safe, but in general it is undecidable if a GHC program is safe.

# GUARDED HORN CLAUSES

### *Language*

**dialect:** committed choice language
**scope:** Horn clauses
**calculus:** resolution
**computational model:** AND/OR tree
**parallelism:** mainly AND-parallelism
**control of parallelism:** implicit through guards and semantical rules

### *Abstract Machine*

as described in (Kimura and Chikayama, 1987)
**type:** adaptation of Warren's abstract machine
**topology:** AND/OR tree of processes (shōen)
**communication mechanism:** (active and passive) unification
**task management:** goals can be ready, suspended, or currently under evaluation; scheduling and suspension of processes via process records and pragmas (priorities)
**parallelism:** goal reductions as independent processes
**control of parallelism:** goal forks, process scheduling via a ready-goal stack and suspension

### *Implementation*

based on (Goto et al., 1988);
**implementation language:** KL1-B
**operating system:** PIMOS
**host:** PSI
**node:** processing elements built around dedicated tagged microprogrammable CPU
**topology:** shared memory for clusters of 8 PEs, multiple hypercube network for 16 clusters
**communication:** shared data within clusters, message passing between clusters

Table 3.7: GUARDED HORN CLAUSES

# Parlog

PARLOG belongs to the same family of concurrent logic programming languages as GHC and Concurrent PROLOG. The main difference between PARLOG (Clark and Gregory, 1986; Foster and Taylor, 1987; Gregory et al., 1989) and GHC lies in the treatment of variable bindings during clause tries. PARLOG uses syntactic mode declarations in the procedure definition, specifying which arguments are to be used as input or output parameters, whereas GHC has semantic rules to make such differentiations; both approaches allow efficient implementation by compiled unification. Concurrent PROLOG uses read-only variables in the procedure call for this purpose, requiring either global program analysis or run-time checks.

PARLOG also contains sequential conjunction (&) and clause separation (;) operators for explicit restriction of parallelism. PARLOG, as GHC, does not permit the binding of goal variables during a clause try; only non-modifying tests are allowed. This leads to a simple two-phase execution model, where input data are tested during the clause try and then output data are created during the evaluation of the clause body. Whereas full PARLOG permits the use of arbitrary goals in the guard part, *Flat* PARLOG only allows pre-defined simple operations. The result is a simple computational model where the state of a computation is represented by a 'flat' pool of processes with a single environment and no hierarchical connections.

Most PARLOG programs are also Flat PARLOG programs, and the rest can be transformed into the flat version; Flat PARLOG is essentially equivalent to Flat GHC (Takeuchi and Furukawa, 1986).

As with the other concurrent logic programming languages, the intended application area of PARLOG comprises also systems programming. This requires effective metacontrol mechanisms to manage resource allocation and exception handling of subcomputations, provided through the PARLOG control metacall together with appropriate scheduling structures for prioritized execution of metalevel programs.

# Parlog

---

*Language*

**dialect:** committed choice language

**scope:** Horn clauses

**calculus:** resolution

**computational model:** AND/OR trees for full Parlog, process pool for Flat Parlog

**parallelism:** AND-parallelism, restricted OR-parallelism during guard evaluation, stream parallelism

**control of parallelism:** mainly implicit and through data-flow constraints for synchronization; also sequential conjunction (&) and clause separation operators (; )

*Abstract Machine*

Flat Parlog machine (Foster and Taylor, 1987)

**type:** derived from flat Parlog

**topology:** pool of processes

**communication mechanism:** unification

**task management:** heap of Parlog data structures and process records

**parallelism:** independent processes

**control of parallelism:** process scheduling and suspension

SPM, an abstract machine for the execution of Parlog on uniprocessors, is described in (Gregory et al., 1989); it is based on an AND/OR tree model, viewing the evaluation of a Parlog program as traversal and modification of this tree.

*Implementation*

on a number of parallel systems (Sequent Balance, ALICE, DACTL)

**implementation language:** C, DACTL

**operating system:** UNIX

**host:** e.g. SUN

**node:** microprocessors

**topology:** bus (Sequent)

**communication:** shared memory, packets

Table 3.8: Parlog

# PARTHEO

PARTHEO (Schumann and Letz, 1990; Schumann and Letz, 1989; Kurfeß et al., 1989) is a sound and complete theorem prover for full first order logic developed within the framework of the European ES-PRIT 415 project. It is derived from the sequential SETHEO which features a performance comparable to semi-compiled PROLOG systems. As underlying proof calculus model elimination is used, and in its present implementation PARTHEO exploits mainly OR-parallelism in different modes (Ertel et al., 1989; Bayerl et al., 1989).

The core input language of PARTHEO is clausal form; a transformation module is available, however, which converts formulae in standard first order logic notation (i.e. with quantifiers and logical operators) into skolemized clause normal form. The input formula may be subject to preprocessing, which aims at a reduction of the size of the formula and the corresponding search tree, e.g. through tautology reduction, subsumption reduction, restricted unit resolution, and other reduction techniques (Letz et al., 1990b).

The formula is then compiled into abstract machine code, optimized with respect to clause ordering, and loaded onto the network for execution on the PARTHEO Abstract Machine, which is a modified and extended Warren Abstract Machine.

The underlying proof calculus for PARTHEO is model elimination, which can be seen as a variant of the connection method. Model elimination exploits the basic idea of the connection method, to check all paths through the formula for tautology[3], by constructing a tree or tableau labeled with literals from the formula. The different solutions investigated through OR-parallelism then can be viewed as a tree of tableaux. Different processors work on different tableaux, where they either find a solution, fail or create new tableaux. Work is distributed through a task-stealing model: a processor requests a task form a neighbor if its local task store is empty. A task does not comprise the whole subtree, but only information about its construction, which is used to rebuild the subtree.

Its performance is 17 KLIPS for one Transputer node (T800) with a speedup of about $0.55 \times$ *number of nodes* for representative examples (Schumann and Letz, 1990).

---

[3] or contradiction in the refutation approach

<table>
<tr><td align="center">

# PARTHEO

</td></tr>
</table>

### *Language*

**dialect:** LOP (full first order logic language with a variety of notations)
**scope:** full first order predicate logic
**calculus:** connection method / model elimination
**computational model:** AND/OR-tree similar to resolution; additional reduction steps
**parallelism:** OR-parallelism
**control of parallelism:** mainly implicit

### *Abstract Machine*

**type:** parallelized and enhanced adaptation of Warren's abstract machine
**topology:** distributed pool of independent, OR-parallel tasks
**communication mechanism:** task and control messages
**task management:** reproof mode instead of complicated environments
**parallelism:** independent tasks / processes
**control of parallelism:** decentralized task attraction by idle processors

### *Implementation*

**implementation language:** C
**operating system:** UNIX, MS-DOS
**host:** SUN 386i
**node:** Transputer
**topology:** various (hypercube, torus, doubly twisted ring)
**communication:** message passing

Table 3.9: PARTHEO

## PEPSys

Aiming at the design and evaluation of a multiprocessor system for large scale parallel logic applications, PEPSYS (Parallel ECRC PROLOG System) has been developed at the ECRC (European Computer Industry Research Center), Munich, since 1984. As an "all-solutions" system, PEPSYS offers simple and consistent extensions of PROLOG while maintaining its declarative semantics. It is able to handle a mixture of OR-parallel, AND-parallel and sequential execution under explicit user control. The model is aimed at easy conversion and low overhead for parallelism. PEPSYS comprises a high-level parallel logic language, a parallel computational model, an abstract machine with the required execution mechanisms, and has been realized as an implementation on a commercial multiprocessor and as a simulation for cluster-based multiprocessor systems (Ratcliffe and Robert, 1986; Baron et al., 1988; Chassin de Kergommeaux et al., 1988).

An evaluation of the PEPSYS model has been done by a performance analysis based on

**Abstract Analysis:** PEPSYS programs are transformed into conventional PROLOG programs with additional predicates; the execution of such a transformed program produces statistical results ( number of inferences, length of processes, maximal speedup), a graphical depiction of the potential parallelism at a specific point, and a trace of the evaluation.

**Simulation:** An event-driven simulator for clusters of shared-memory machines has been used to obtain more realistic performance measurements for large systems.

**Implementation:** A real parallel system has been used as the most reliable basis for performance measurements.

As a result, a tight correlation between implementation and simulation, and a broader, but still obvious correlation between all three evaluation systems could be found, sufficient for a reasonable extrapolation of the implementation measurements towards large scale systems based on the simulation and abstract analysis.

# PEPSYS

### *Language*

**dialect:** PEPSYS extension of PROLOG

**scope:** Horn clauses

**calculus:** resolution

**computational model:** serial or (OR- and independent AND-) parallel modules, coarse-grained processes; management of variable bindings with time-stamped OR-branch levels and hash windows; sequential backtracking, some precautions maintaining completeness for mixed OR-parallel, AND-parallel and sequential computations

**parallelism:** OR-parallelism, independent (restricted) AND-parallelism, induced AND-parallelism (as a consequence of OR-parallelism)

**control of parallelism:** explicitly by the user (serial and parallel modules, predicate property declarations, special operators)

### *Abstract Machine*

**type:** extension of WAM for variable binding management, OR- / AND-parallel execution, backtracking

**topology:** (clusters of) bus-connected processing elements

**communication mechanism:** shared memory

**task management:** UNIX processes

**parallelism:** processes

**control of parallelism:** create / kill processes, inter-process synchronization (continuation mechanism for next subgoal, backtracking)

### *Implementation*

**implementation language:** C

**operating system:** UNIX

**node:** NS 32032

**topology:** bus

**communication:** shared memory

Table 3.10: PEPSYS

# PIM

The design and implementation of parallel inference machines has been a major goal in the Japanese Fifth Generation Computer Systems project (Goto et al., 1988; Uchida et al., 1988). The aim is to realize very high performance for the evaluation of logic programs by a system based on a logic programming framework for the whole system. This framework comprises a whole family of languages on various levels of abstraction, all based on a logic paradigm with GHC as its main representative.

The execution of GHC programs relies on *shōens*, which define large-grain computational units, roughly corresponding to a task or a job. Shōens realize a metaprogramming capability into KL1, the kernel language of the PIM system, which has been used for the implementation of PIMOS, the PIM operating system. They are responsible for execution control, resource management and exception handling, and define the hierarchical structure of a program. Through metaprogramming, a clear separation is achieved between operating system programs (object level) and user programs (meta-level).

Scheduling and distribution of object level goals is of extreme importance for efficient execution; PIMOS uses a non-busy waiting goal scheduling mechanism with priorities, and has two kinds of goal distribution schemes: one for tightly coupled processors (within PIM clusters), and one for loosely coupled processors (between PIM clusters).

For efficient program execution, dedicated architectures have been designed. PIM-p is a based on single-board processing elements eight of which together with a shared memory are interconnected via a bus to form a cluster; each processing element has a network interface unit for inter-cluster access. Execution within a processing element is based on a four-stage hardware pipeline, and a special internal memory is used to enhance the evaluation of macro instructions

A performance of 10-20 MLIPS is envisaged for a system with 100 processing elements, and 200-500 KLIPS for a single PE.

# PIM

## *Language*

**dialect:** KL1 (based on GHC), KL1-B as abstract instruction set

**scope:** Horn clause logic

**calculus:** resolution

**computational model:** GHC / Shōen; guards control execution of a clause.

**parallelism:** AND-parallelism

**control of parallelism:** shōen, guards, pragmas assign goals to PEs

## *Abstract Machine*

**type:** similar to a parallelized WAM, defined by the KL1-B abstract machine instruction set

**topology:** hierarchy of clusters

**communication mechanism:** shared memory within a cluster, message-passing between clusters

**task management:** according to the shōen structure; ready-goal stack with priorities in a cluster, pragmas among different clusters

**parallelism:** AND parallelism, streams

**control of parallelism:** shōen, guards, pragmas to assign goals to processors, depth-first scheduling, priority of goals, suspension

## *Implementation*

**implementation language:** KL1-C; system description language with modularization, macro-expansion, built-in predicates

**operating system:** PIMOS; described in KL1-C, self-contained, single system for the user, independent of the underlying architecture

**host:** PSI-II (sequential inference machine)

**node:** experimental systems as networks of PSIs (Multi-PSI-V1, -V2); dedicated nodes (PIM-p) in development

**topology:** tightly coupled clusters of eight PEs and a shared memory, multiple hypercube network for inter-cluster communication

**communication:** shared memory and message passing

Table 3.11: PIM

## POPE

The approach pursued in POPE (Parallel-Operating PROLOG Engine) (Beer and Giloi, 1987; Beer, 1989) is somewhat different in that it does not rely on any of the classical forms of parallelism in logic programs (i.e. AND-, OR-parallelism), but on a parallel execution of standard sequential PROLOG, including critical features like dynamic assert / retract, or the cut operator. The architecture consists of a pipeline of unification processors to be used as coprocessor for conventional systems. The single processors are micro-programmable and have been designed to execute abstract PROLOG machine instructions. The processors may have local memory, but have access to a global system memory. Communication between adjacent stages of the pipeline is handled through pipeline buffers, with hardware semaphores to synchronize the (uni-directional) exchange of data.

During execution, a processor sends the entry address of a clause to the next processor, followed by the arguments of the current goal. His successor gets the clause head through the entry address and tries to unify the arguments of the current goal with the clause head; before unification starts, however, the entry address of the first subgoal is sent to the third processor, followed by the arguments as they become available. If an argument is not available yet, a processor automatically delays execution until the argument becomes available. Possible binding conflicts are avoided by controlling the access to the trail stack through the hardware semaphores of the pipeline buffers.

The information transmitted between the pipeline stages contains all the state information of a particular process. This allows the concurrent execution of multiple PROLOG programs in the pipeline.

The simulation results show that even programs without 'classical' forms of parallelism can be executed efficiently; in the examples discussed, however, a pipeline consisting of only three processing elements was sufficient. Access to the global shared memory also proves to be a limiting factor.

# POPE

*Language*

**dialect:** full standard PROLOG

**scope:** Horn Clauses

**calculus:** resolution

**computational model:** pipelined resolution of goals and their arguments

**parallelism:** not visible on the language level

**control of parallelism:**

*Abstract Machine*

**type:** collection of modified WAMs

**topology:** pipeline

**communication mechanism:** shared memory for PROLOG stacks, pipeline buffers for sub-tasks (goal + arguments, corresponding clause head)

**task management:** transfer of sub-tasks to the right neighbor

**parallelism:** overlapping execution of PROLOG programs

**control of parallelism:** program structure, data flow through the pipeline, hardware semaphores

*Implementation*

**implementation language:** microinstructions implemented as C routines

**node:** microprogrammable nodes designed to execute an abstract PROLOG instruction set

**topology:** circular pipeline

**communication:** shared memory, pipeline buffers, hardware semaphores

Table 3.12: POPE

## 3.2   Conclusions

Research is very active in this area, and the first commercial products are already emerging (CSPROLOG (Futo and Kacsuk, 1989), STRAND88 (Foster and Taylor, 1989)). Two main directions can be identified, one concentrating on a parallel execution of PROLOG as it is ("All-Solutions" Languages), aiming mainly at increased performance, the other giving up compatibility with PROLOG for an appropriate handling of parallelism issues like communication and synchronization ("Committed Choice" Languages).

The first approach uses mainly OR-parallelism to increase performance, but faces problems with the required management of hierarchical variable bindings implied by the AND/OR tree based computational model. Some proposals include the usage of (restricted) AND-parallelism (PEPSYS (Baron et al., 1988), DeltaPROLOG (Cunha et al., 1989)), and ANDORRA (Haridi and Brand, 1988) provides a reasonable integration of both concepts. The support for *describing* parallel systems essentially is the same as in PROLOG, requiring additional constructs like modules or objects on top of it. A parallel logic system featuring both description of parallel systems as well as parallel evaluation of logic programs with a smooth transition from the specification to the implementation is severely restricted by the adaptation of PROLOG's operational semantics with its dependencies on the evaluation order (e.g. `cut`), and the inclusion of constructs with questionable logical basis (e.g. `assert`).

AND-parallelism is the main basis for the second approach; its operational semantics facilitates parallel evaluation, and it is also used with considerable success to model parallel systems. The description of such a parallel system, however, must be tailored to comply with the evaluation model of the language, which is appropriate for a restricted class of problems. Proposals to integrate more general concepts (like objects) have been made for Concurrent PROLOG (Shapiro and Takeuchi, 1983) and for GHC / KL1 (Ohki et al., 1987), and for concurrent logic languages in general (Kahn, 1986).

The third group discussed in this section, "Pure Logic" languages, may have both high expressiveness for specification purposes and considerable potential for parallel execution, but still are in initial stages with respect to usability and efficiency.

# Chapter 4

# Parallelism in Logic

This chapter deals with the investigation of parallelism in logic, and its exploitation. In the prelude, it offers some *variations on parallelism*, outlining general models, or *categories*, of parallel computation. Then, after a brief overview of the different categories of parallelism to be found in logic, each of these categories is discussed in more detail. Typically, such a discussion provides a problem specification, the design and formal description of a solution, considerations about the correctness and complexity of the solution, its potential for internal parallelization, and possible refinements. The important aspects of the discussion are comprised in an overview table, enhanced by criteria relevant for an implementation. In addition, possible combinations of categories are investigated. The chapter concludes with the proposal of some favorable combinations of categories; of particular importance for good combinations are the underlying calculus, the computational model applied, and the execution vehicle envisaged.

# 4.1   Chapter Organization

Since this chapter represents a central part of the book and contains a description of complex and interrelated features with many facets, its organization will be explained in more detail. As kind of a prelude, the *variations on parallelism* illuminate some general categories of parallelism without particular stress on the interrelation between logic and parallelism. In the next section, a short *overview* is given about the different categories of parallelism eminent in logic programming, together with some considerations towards a taxonomy of the different categories. The actual structuring of the main part of the chapter, as mirrored in the sequence of the single sections on the *different categories*, is derived from a more practical aspect, namely the granularity of the computational task within a category. Whereas this ordering sometimes does not represent conceptual interdependencies very well, it has the advantage that single categories do not appear in different places, and the ordering can be obtained in a straightforward way. Some of the categories of parallelism discussed in this chapter can be clearly separated from another; others, however, are somehow intertwined, and could surely be arranged in a different order or regarded according to different aspects.

The investigation of a particular category of parallelism in logic is done according to the following structure of the corresponding section:

**Description of the Category**
- in the *problem specification*, relevant aspects of the problem are identified;
- the *design of a solution* describes the underlying idea for solving the problem and extracts its important properties;
- a UNITY *program* formally specifies the solution, and provides a sound basis for
- considerations about the *correctness* and *complexity* of the solution;
- the potential for *internal parallelization* within the category itself is explored;
- *possible refinements* are proposed, often directed at a more efficient computation;
- in an *overview table*, the basic data structures and operations for representation, computation, and communication purposes of a particular category are summarized.

### Combinations of Categories

- the category itself, invoked in multiple instances side by side, or in a nested (recursive) way;
- combinations of this category and previously described ones.

### Conclusions

- the section describing a particular category is ended by conclusions on its practical usability.

In a few cases, some categories listed explicitly in the overview are not treated separately but grouped together since their important features are closely related.

## 4.2   Variations on Parallelism

The notion of *parallelism* is used for a class of executional models where more than one tasks or processes are handled simultaneously; some aspects that might be used to distinguish between different variations of parallelism are illuminated in this section, describing inherently parallel features of models to be exploited concurrently. In the following we refer to the instances of this class as various *categories* of parallelism.

## Multiple Tasks: Treating Different Problems Simultaneously

> *"Dis, Astérix, qu'est-ce qu'on fait*
> *de ces cent romains-là?"*
> Obélix

The treatment of more than one task in a way that looks simultaneous to the user is a common feature of conventional computer systems (comprising multi-tasking, multi-programming, and multi-user systems) and is beginning to be available even for personal systems. The thread of execution, however, is regarded as (multi-)sequential or quasi-parallel since there is only one sequence of instructions being executed. The extension of multi-tasking to parallel systems is quite straightforward and in some aspects even simpler than the complex task switching mechanisms for sequential systems. Tasks usually are relatively self-contained, without a need for complex synchronization mechanisms. Problems similar to the ones encountered in sequential systems have to be envisaged for the common use of shared resources (code, data, memory, special units, peripheral devices) and in overload situations; in general, the administrative overhead in the parallel case might be about the same or lower than in the sequential one, and simultaneous execution of multiple tasks is conceptually easier in parallel systems.

## Partitioning: Dividing a Task into Pieces

*"Divide et Impera"*
A Roman Emperor

Many tasks to be performed by a computer system consist of a set of subtasks with only few interrelations. These interrelations can be based on *data interdependencies*, or on the *flow of control* defined via an algorithm and ist execution. The range of partitioning reaches from a very high granularity (a task accomplished by a collection of processes) over medium size (numerical computations are executed by a floating point coprocessor) to a very low one (horizontal microcoding: independent parts of a machine instruction are executed simultaneously in different functional units of the CPU)[1].

## Competition: Different Ways to a Solution

*"Many ways lead to Rome"*
A Roman Stroller

The art of programming a computer has two aspects that sometimes diverge and sometimes coincide: one is the clarity of the underlying conceptual model, the other is the efficiency of the actual implementation. In sequential systems, the choice is usually for the efficiency point; parallel systems can offer the exploration of more than one paths to find the solution to a given problem. They may be derived from different computational models (procedural, object-oriented, functional, logic), different algorithms (quicksort, bubblesort), different evaluation strategies (depth-first vs. breadth-first search, recursion vs. iteration), or even different underlying hardware (e.g. floating point unit,

---

[1] This approach has been pursued on its lower and medium levels of granularity from the very beginning of computer engineering.

vector processor). One could argue that this is in principle a waste of processing power because redundant computations are performed; however, it is generally not known in advance which approach is the best one. So it may be better to pursue different ones simultaneously, thus increasing the probability for a fast solution, instead of sticking to one with the risk of a bad choice, resulting in poor performance or even failure. The depth-first, left-to-right strategy applied in PROLOG implementations is a good example: consider the case that the system follows the leftmost branch of the search tree endlessly while the solution can be found easily (i.e. close to the root) in another branch.

## Precision: Distinct Levels of Accuracy

> *"Genauigkeit hat ihren Preis"*
> G.E. Nau

The quality of a solution very often depends on the effort to be invested; in some cases a simple estimation can be sufficient, whereas in others an exact result is essential. Superficial methods may be used to check out the usability of a result, or the feasibility of an approach, and more detailed ones to compute the precise outcome. Additionally, the information gained on more superficial levels can be used to guide the detailed evaluation process, and important details help to improve the quality of an estimation. Investigating a task in parallel on distinct levels of accuracy can result in synergetic effects expediting the evaluation process, and avoid being stuck with non-optimal solutions (local maxima/minima), but it also carries the risks of superfluous work.

## Multiple Solutions: One Result is not Enough

> *"'Twas brillig, and the slithy toves*
> *Did gyre and gimble in the wabe;*
> *All mimsy were the borogoves,*
> *And the mome raths outgrabe."*
> Lewis Carroll

The completion of a task sometimes requires finding out more than one data item satisfying the specifications for a solution; a typical example is a query in a data base. In sequential systems this is often done via iteration, whereas in parallel ones slightly varying tasks could be treated by different processing elements. This implies either additional control information or some agreement on who tries to find which solution. In the case that different solutions can be identified and separated easily (e.g. by an additional parameter), this category can be subsumed by 'Multiple Tasks' and / or 'Partitioning'.

## Problem Representation: Another Dimension

> *"I call our world Flatland, ...*
> *Imagine a vast sheet of paper ... "*
> Edwin A. Abbott

The way the human brain tackles problems usually is highly influenced by the three-dimensional perception delivered by the organs of sense; likewise is the way we program a computer system to process some task influenced by the conventional sequential program execution paradigm. In many problems there is a high potential for efficiency gain (sometimes leading to feasibility) by choosing an appropriate number of dimensions for representation. This fact is well-known in mathematics (esp. analytical geometry and graph theory), is sometimes applied in conventional, sequential and vector processing systems, and may show a lot more favorable effects in combination with parallel systems. In the particular case of logic, the representation of a formula in an alternative way (e.g. as matrix) leads to easier understanding and better computational treatment. The notion of representation as used in this paragraph not only refers to the storage of data in a computer system, but also to the description of execution patterns through algorithms, or to communication activities in a computer system.

## Cooperation: Exchanging Useful Information

> *"Tous pour un,*
> *un pour tous"*
> Alexandre Dumas

Especially in combination with the *competition* paradigm, some processing element might yield information (partial results, information about the prospects of a particular strategy) which can be very useful for other processing elements in their trial to solve a problem. Cooperation can take place on the object level (e.g. by the exchange of already computed data, or by splitting up work among specialized units), or on a meta-level by providing or deriving information on the computation process itself. The problem here is to judge the usefulness of information; in general, this requires knowledge about the computation process, and about the status of the other processing elements. In non-centralized system the lack of accurate overall status information is a fundamental issue making some methods obsolete because they explicitly or implicitly rely on the availability of overall status information.

### Distributed Representation: A Holistic Approach

*"The whole is simpler*
*than the sum of its parts"*
**Willard Gibbs**

A paradigm of computation quite different from von Neumann's is based on the non-local representation and processing of data; there are some indications that this paradigm has certain similarities to the way the human brain works, and after a precipitate cut in the late sixties a lot of research effort is being carried on lately in that field. In principle, distributed computation could be subsumed by the competition category since it can be viewed as a different way to find a solution to a problem. It is mentioned here explicitly since it offers a source of parallelism orthogonal to the more conventional ones described above and looks quite promising for some classes of problems where conventional paradigms of computation show weaknesses.

## 4.3   Overview

Theorem proving and logic programming offer a large potential of parallelism not only restricted to the AND/OR-parallelism usually investigated in attempts to parallelize PROLOG. In this section, different sources of parallelism are identified during the process of developing a logic program from an initial problem description to an execution of a corresponding program on a particular (parallel) computer system. The phases during program development are listed below, together with the tools or methods used in a certain phase, and the potential categories of parallelism relevant for that phase.

**Specification:** *Description of a problem and design of a solution*
- Tool / Method: Paradigm of computation (e.g. logical, functional, procedural, object-oriented)
- Parallelism: Specifications according to alternative paradigms, problem structure (modularity), different solution ideas.

**Evaluation:** *Validation of the solution*
- Tool / Method: Algorithm, computational model, abstract machine
- Parallelism: Different logical calculi, different strategies, reductions, calculus-specific kinds of parallelism (routes, spanning sets).

**Execution:** *Computation of the solution*
- Tool / Method: Implementation language, execution platform
- Parallelism: AND/OR-parallelism, term parallelism, cooperation, multitasking, stream parallelism

The actual treatment of the different categories is organized in a slightly different way, starting from large-grain features like multitasking and modularity, over medium-grain ones like reductions, OR- and AND-parallelism, to variations on term parallelism, and distributed representation models. Figure 4.1 depicts the relation of the different categories with respect to their granularity on one axis, and their independence on the other. The *granularity* is characterized by the pieces of a formula treated within one category; on the top level are *independent formulae* (as in multitasking) or relatively independent parts of one formula (modularity); the next level comprises activities which require the *whole* formula for performing computations like competitive evaluations, reductions of the search space, or the computation of spanning sets. Dealing with *clauses* as items are certain reductions again, and OR-parallelism, whereas AND-parallelism is based on *literals*. Activities on the term level are concerned with unification and related issues like occur check, and this level is also termed *unification* or *term parallelism*. On a level below the single *atoms* of a formula appear concepts like distributed representation, where there is no one-to-one correspondence between a symbol and its internal representation (e.g. in a memory cell), but rather a distribution of the information for one symbol over a number of memory cells. *Independence* describes the necessity of information exchange between single computations of one and the same category. The treatment of different formulae in multitasking, or the computation of different spanning sets for one formula, for example, are activities which require almost no interaction or exchange of information between the single tasks. On the other hand, the occur check, or computations based on the distributed representation paradigm, rely totally on interactions, and the single computational units of one category are strongly dependent on each other in order to achieve the goal of their computation. The category of *cooperation* represents a special case here; it can be applied on and among various levels of granularity, which is depicted by the dashed box surrounding all the categories appearing in Figure 4.1.

**Multitasking:** *Executing more than one tasks (logic programs) simultaneously.*

It is usually handled by the operating system, not the user program, and involves activities like task administration, resource management, process management (creation, synchronization, killing).

**Modularity:** *Composition of large programs.*

Often systems to be described can be modeled in terms of relatively independent modules; in general, this requires *higher order logic* since it incorporates arguing about properties of predicates. Usually ad-hoc approaches are pursued with a simple syntax (`begin module - end module`) and questionable semantics. It is very important for the description of parallel systems via logic programs.

**Precision:** *Different unification mechanisms.*

Unification can be performed with various restrictions, resulting in different levels of accuracy, abstraction. The range can go from no unification (i.e. propositional logic), over term unification, directed unification, unification without occur-check, and weak unification to full unification. The problem here is that (full) unification is very expensive in terms of memory and execution time.

**Competition:** *Different evaluation strategies.*

Execution of logic programs can be based on evaluation mechanisms with various underlying *calculi*: resolution (as in PROLOG), connection method, tableau calculus, and others. In addition, different *search strategies* to traverse the proof tree can be applied: depth-first/breadth-first, bottom-up/top-down, or 'intelligent' strategies using domain-dependent knowledge, information gathered during the proof, or meta-knowledge (learning from previous proofs, implicitly programmed by the system designer).

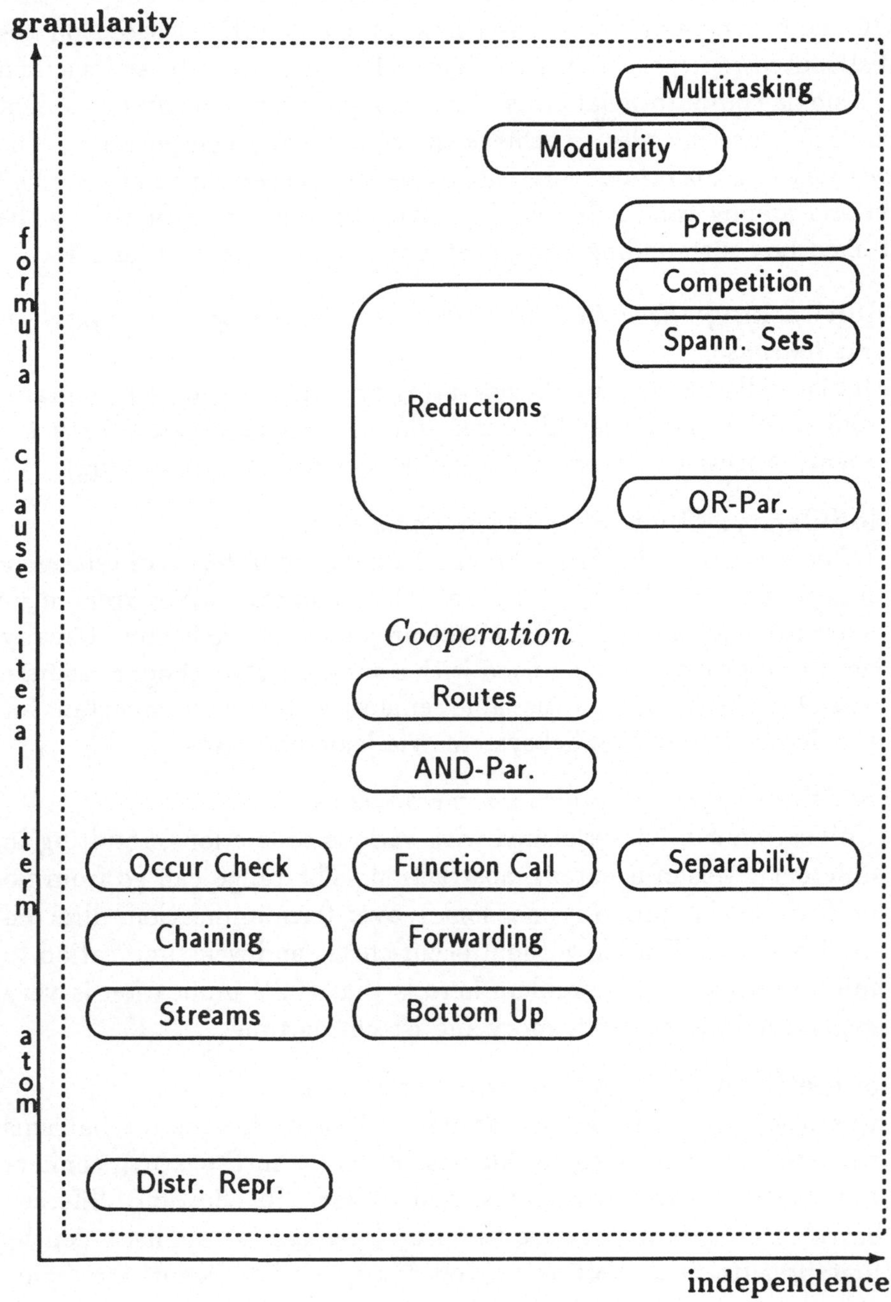


Figure 4.1: The Space of Parallelism in Logic

**Spanning Sets:** *Computation of alternative solutions.*
Statical analysis of the formula leads to alternative sets of clauses, describing different possible solutions. Its result is the same as OR-parallelism, but derived via self-contained proof attempts instead of the partial proofs spawned at appropriate branches in OR-parallelism.

**Reductions:** *Transformations to trim down the search space.*
They can be categorized into two classes: preserving *equivalence*, (like the first four in the following list) or *satisfiability* (like the rest).

- Tautology reduction: removes a clause $c$ containing a literal $L$ and its negation $\neg L$.

- Subsumption: a 'stronger' (more general) part of the program subsumes a 'weaker' one.

- Factorization: a clause $c$ subsumes a proper subset $c'$ of itself.

- Restricted unit resolution: (more complicated).

- Pure literals: literals without a unification partner.

- Isolated connections: pairs of literals without 'connections' to the rest of the formula.

The *scope* of reductions can be global, concerning the whole formula, (like the first four examples), or local on parts of the program only (single clauses, pairs or sets of clauses). These reductions, especially combinations of them, very often drastically reduce the search space to be traversed (snowball effect).

**OR-Parallelism:** *Pursuing diverse solutions.*
OR-parallelism occurs in the resolution calculus when there is more than one unification partner (clause head) for a literal. It is exploited during execution time, and must include precautions for multiple variable bindings (cactus stacks, hash windows).

**Cooperation:** *Beneficial interaction between parallel activities.*
The main goal here is to make potentially useful information available for others. It can be applied on various levels:

- Meta-level: evaluation mechanism used (calculus), strategies applied, proof schemata;

- Proof level: information about the proof process, e.g. 'dead' or 'prolific' branches ;

- Evaluation level: progress of the work, current activities;

- Data level: (sub-) results, e.g. lemmata.

At the moment, cooperation is more art than science; mostly it is implicit in the implementation.

**Routes:** *Paths plastered with connections.*
Again by a statical analysis of the formula the parts of paths through a matrix which are associated through connections can be identified. The terms of the literals of a route are combined into a dag, and then unified in one run, instead of step by step as in the resolution calculus.

**AND-Parallelism:** *Dividing the clause body.*
The literals of the clause body (sub-goals) are evaluated in parallel. Synchronization is necessary because all sub-goals have to be solved and shared variables can occur. This mechanism can be used for modeling parallel systems via logic programs (guards, read-only variables) Implementations are usually restricted to the evaluation of independent subgoals.

**Occur Check:** *Check for cyclic substitutions.*
The substitution of a term by a subterm of itself leads to a cycle; in many PROLOG implementations no occur check is performed due to reasons of efficiency. It can be done independent of the actual proof process and corresponds to searching a graph for cycles.

**Function Call:** *Unification and evaluation of function occurrences.*
Terms with function symbols (in corresponding positions) can only be unified if the function symbols are identical; thus, different occurrences of the same function may be unified in advance and independent of the proof structure. This can cause superfluous work (performing unifications and function evaluations not needed in the actual proof process), but may be useful to avoid idle times.

**Separability:** *Partitioning term structures.*
In some cases, pairs of huge term structures have to be unified ('dag' approach); these terms can be partitioned into (relatively) independent sub-terms. It can be prepared at compile time by an analysis of the term structures involved. Problems occur with shared variables. It is fine grain parallelism, similar to modularity on the program level.

**Forwarding:** *Passing on the work.*
Two terms are unified in a top-down approach, starting from the roots. This incremental unification results in extremely fine grain parallelism. It can be prepared statically and might be suitable for massively parallel systems like the Connection Machine.

**Bottom-Up:** *Passing on the results*
This is similar to forwarding, but starting unification from the atoms (leaf nodes) instead of the root nodes. The information about the nodes to be unified initially can be derived using a forward-chaining evaluation mechanism. It is especially promising for programs with many facts (constants).

**Chaining:** *Accumulating subterms to be unified.*
It is derived from a static analysis of the program. In contrast to PROLOG's stepwise evaluation by unifying *pairs of terms*, chaining consists of unifying *'chains'* or *sets of subterms* in parallel. A problem lies in the treatment of literals at 'intersection' points of chains which should be evaluated independently.

**Streams:** *Handling incomplete data structures.*
In many programs performing symbolic computations (especially on lists), data structures tend to be quite complex; sometimes it is possible to already perform a second operation on a part of the structure before the original computation on that structure is finished.

**Distributed Representation:** *Spreading information items over a large number of nodes.*
Logic and reasoning may be pursued with a variety of approaches on different levels of complexity. A *reasoning system* can calculate the plausibility or consistency of a collection of statements (logical program) in a way similar to energy minimization. Neural models are able provide support for *heuristic decisions*, (Suttner, 1989), or can realize a device for term *unification*. (Kurfeß, 1991; Hölldobler, 1990b; Stolcke, 1988). A *proof checker* can validate the consistency of variable bindings. A full *proof system* based on neural models could make use of their inherent capabilities of learning, abstraction and generalization. This approach has some problems with standard logic based on binary values; it may offer advantages for non-standard logic (e.g. fuzzy logic), and for the description of real-world problems with their inconsistencies, incompleteness and temporal change.

The exploitation and combination of these different kinds of parallelism must be tackled quite carefully; otherwise the result may be severe obstruction of the parallel system used to evaluate a logical formula.

# 4.4   Multitasking

The ability to handle more than one task simultaneously, or at least quasi-simultaneously in the sense that to the user it looks like multiple tasks are evaluated simultaneously, is a very essential feature of computer systems. It is indispensable to systems which are shared by some users, and even begins to permeate into the area of single-user systems like personal computers. In our context of logic programming, *multitasking* is defined as the simultaneous evaluation of a set of logic programs (or formulae); it encloses the terms *multiprocessing* (running more than one process in the system), *multiprogramming* (evaluating more than one program), as well as *multi-user* features (more than one user can be handled simultaneously). These logic programs are independent: they do not share any data. Since the programs are independent, dependencies during the evaluation may only be caused by the operating system due to reasons like lack of resources (processors, memory space) or increase of efficiency (code sharing). These problems will not be handled in this section; some indications for their treatment in parallel systems can be found in the chapter on architecture.

### Problem Specification

For a given set of formulae $\mathcal{F} = \{F_1, \ldots, F_n\}$: try to find a proof $P_i$ for each formula $F_i \in \mathcal{F}$. A *task* consists of the evaluation of a single formula.

### Design of a Solution

From the above specification, a solution can be derived quite easily. As a starting point a generic task execution mechanism TASK-EXEC($\mathcal{TA}$) is used. $\mathcal{TA}$ is a set of tasks $\{TA_1, \ldots, TA_n\}$, where a task $TA_i$ consists of a triple ($input_i$, $computation_i$, $output_i$), and the execution of a task is achieved by applying $computation_i$ to $input_i$ resulting in $output_i$. A fixed point for the whole program is reached if and as soon as every task $TA_i$ reaches a fixed point. The program TASK-EXEC (see Figure 4.2) provides a formal UNITY specification of the solution derived here.

### Refinements: Logic Programs as Tasks

For each formula $F_i \in \mathcal{F}$, a task $TA_i$ is defined as the formula together with an evaluation mechanism EVAL, and a proof $P_i$: $TA_i =$

**Program TASK-EXEC**
**parameters**

| name | type | description |
|------|------|-------------|
| $TA$ | set of tasks | tasks to be executed |
| $TA_i$ | task | single task |
| $input_i$ | set of data items | input data of $TA_i$ |
| $output_i$ | set of data items | output data of $TA_i$ |
| $computation_i$ | set of statements | program of $TA_i$ |

**pre-conditions**

$TA \neq \{\}$ — there are some tasks

$input_i \neq \{\}$ — input data of $TA_i$ not empty

$output_i = \{\}$ — output data of $TA_i$ are empty

$computation_i \neq \{\}$ — program of $TA_i$ is not empty

**invariants**

$TA_i = (input_i, computation_i, output_i)$ — task components

**computation**

$$\big[\!\big] \quad \langle \quad i : TA_i \in TA :: \qquad \text{for all tasks}$$
$$\qquad \big[\!\big] \; result_i \; := \; computation_i \, (input_i) \qquad \text{execute task } TA_i$$
$$\rangle$$

**post-conditions**

$output_i \neq \{\}$ — result found

**end TASK-EXEC**

Figure 4.2: Program TASK-EXEC

$(F_i, \text{EVAL}_i, P_i)$. For a specification of this refined version see Figure 4.3.

## Correctness and Complexity

The correctness of this solution of course depends on the correctness of the single evaluation mechanisms; depending on the input formula, it is also possible in some cases to show that the formula cannot be proven (which can be expressed by returning **not provable** as result of the proof process). The complexity is also determined by the complexity of the tasks under execution.

## Internal Parallelization

The potential for internal parallelization in multitasking also cannot be described without a knowledge about the actual computation to be

**Program TASK-LOGIC**

**parameters**

| name | type | description |
|------|------|-------------|
| $\mathcal{TA}$ | set of tasks | tasks to be executed |
| $TA_i$ | task | single task |
| $F_i$ | formula (set of clauses) | input data of $TA_i$ |
| $P_i$ | proof | output data of $TA_i$ |
| $\text{EVAL}_i$ | program (set of statements) | evaluation mechanism of $TA_i$ |

**pre-conditions**

| | |
|---|---|
| $\mathcal{TA} \neq \{\}$ | there are some tasks |
| $F_i \neq \{\}$ | formula not empty |
| $P_i = \{\}$ | no proof |
| $\text{EVAL}_i \neq \{\}$ | evaluation mechan. not empty |

**invariants**

$$TA_i = (F_i, \text{EVAL}_i, P_i) \qquad \text{task components}$$

**computation**

$$\|\quad \langle \quad i : TA_i \in \mathcal{TA} :: \qquad \text{for all tasks}$$
$$\|\quad P_i \quad := \quad \text{EVAL}_i(F_i) \qquad \text{evaluate formula } F_i$$
$$\langle$$

**post-conditions**

$$P_i \neq \{\} \qquad \text{proof found for } F_i$$

**end TASK-LOGIC**

Figure 4.3: Program TASK-LOGIC

performed during the execution of a task.

Table 4.1 summarizes important features concerning the use of multitasking, especially with respect to the evaluation and execution of logic programs. For this purpose, the data structures and operations essential for computation, representation and communication are summarized.

## Conclusions

The concept of multitasking as known from conventional systems can be applied in parallel systems in a straightforward way. In combination with logic programming, it provides the basis for the independent evaluation of different logic programs on one system. The main problems are encountered in the organization of common resources, and task distribution / load balancing.

| MULTITASKING | | |
|---|---|---|
| | **data structures** | **operations** |
| **compu-tation** | **input data**<br>• set of tasks (logical formulae, programs)<br>• sets of parameters (proof depth, number of inferences, number of copies)<br>**output data**<br>• proof result (truth value)<br>• proof information (substitutions, proof tree, trace)<br>**intermediate data**<br>• task status<br>**evaluation support**<br>• libraries<br>• administrative information: owners, allocation of the tasks, shared resources, system status information | **task execution**<br>• evaluation mechanisms for logical programs (provers)<br>**multitasking system**<br>• user handling<br>• task administration<br>• task allocation<br>• resource administration |
| **repre-sen-tation** | **tasks / programs**<br>• code to be executed by the single provers<br>**local memory**<br>• for the tasks<br>**global memory**<br>• available resources<br>• system topology<br>• current activities | **access services**<br>• main / background memory<br>• (`read / write` instructions)<br>**evaluation support**<br>• memory management (virtual memory)<br>• garbage collection |
| **com-muni-cation** | **implicit**<br>• shared data structures<br>**explicit**<br>• parameters<br>• messages | **implicit:**<br>• mechanisms for the use of shared data structures (`read / write` operations with prevention of simultaneous writes)<br>**explicit:**<br>• parameter passing<br>• message passing |

Table 4.1: Multitasking

# 4.5   Modularity

Structuring programs into modules with only a few, well-defined interdependencies is an indispensable feature for any programming language aiming at the construction of large software packages. The scope of declarations is limited to the module, and interdependencies between modules are usually specified by **import / export** declarations for the use of non-local data structures or procedures. In logic programming languages, the construct to express procedural elements as well as data structures are predicates, and their scope comprises the whole formula; this, of course, presents a severe limitation for the use of logic programming to write large programs. Limiting the scope of predicates means arguing about properties of predicates, which in principle lies beyond first order predicate logic and requires constructs from second or higher order predicate logic.

Some approaches have been proposed to overcome this problem, many of which lack a clear semantics as a sound logical foundation; for a survey see (Barbuti et al., 1987). These approaches can be divided into three classes:

- **Syntactic:** A flat hierarchical organization is imposed onto the program by giving scope rules for predicate symbols, plus some constructs to handle large programs partitioned into modules[2]. The structure lies only in a syntactic view of a single big program (given basically by the symbol tables of the compiler), and every structure disappears at run time.

- **Semantic:** A complex hierarchical organization is defined on programs, and a clear semantics based on logic is used to provide a modularization mechanism for the composition of programs out of pieces. Possibilities to derive a semantics can stem from traditional languages (denotational semantics, fixed points), algebra (defining composition operators for an algebra of programs), or logic (a program is a mapping from an interpretation to an interpretation). Using second order logic constructs can provide a framework for logic programming including hierarchical structures and modularization; the semantics can be defined by viewing modules as sets of second order clauses, and module names as second order predicates. On

---

[2] in PEPSYS (Rapp, 1988; Ratcliffe and Robert, 1986), for example, a module corresponds to a file, and different modules thus must be defined in different files

this basis, a proper integration of procedure definition and procedure call mechanisms can be achieved, where procedures are second order formulae in which universally quantified variables correspond to imported and existentially quantified variables to exported parameters.

- **Meta-programming:** This approach is based on the notion of theories, where multiple theories ('worlds', databases) can be introduced and composed dynamically. Specific inference engines are defined for the theories, containing meta-level operations to load, search, query and update a theory, as well as some tools like debugger, tracer, explainer. Theories and their inference engines conceptually can be seen as knowledge representation mechanisms, using meta-programming for their definition and partial evaluation as implementation technique which reflects a meta-level proof into an equivalent object level proof.

### Problem Specification

The purpose of this investigation is to explore the potential of parallelism in the evaluation of logic programs, concentrated on logic programs as composed of modules. Therefore the main point lies in the *executional* viewpoint, not in the one regarding the features on the language level, and a module is seen as an independent unit of computation with references to some global predicates (as data structures or procedures) defined by something equivalent to `export` / `import` declarations.

Relations between the modules are given implicitly through the shared data structures and their data dependencies. Each of these modules is executed by an evaluation mechanism EVALMOD, taking into account the restrictions imposed by the use of the global predicates in the `import` declaration. If the execution succeeds, it represents a proof for the formula or program specified in the module. According to the philosophy of modularity, the evaluation mechanism EVALMOD can be a standard prover (for partially evaluated modules), a prover lifted to second order logic, or a specific inference engine for a particular theory.

### Design of a Solution

As a first approach a solution is presented for a *flat* modular program consisting of a set of modules without any hierarchical structure; this means that a module may not contain (sub-)modules. A *modular logic*

**Program MODULE-EXEC1**

**parameters**

| *name* | *type* | *initially* | *description* |
|---|---|---|---|
| $\mathcal{MO}$ | set of modules | | modules to be executed |
| $MO_i$ | module (set of clauses) | | single module |
| $P_i$ | proof | | output data of $MO_i$ |
| EVALMOD$_i$ | program | | evaluation mechanism for $MO_i$ |

**pre-conditions**

| | |
|---|---|
| $\mathcal{MO} \neq \{\}$ | there are some modules |
| EVALMOD$_i \neq \{\}$ | evaluation mechanism for $MO_i$ is not empty |

**local variables**

| | | | |
|---|---|---|---|
| $\mathcal{IMP}_i$ | set of predicates | empty | imported predicates |
| $\mathcal{EXP}_i$ | set of predicates | empty | exportable predicates |
| $\mathcal{GLP}$ | set of predicates | empty | global predicates |
| $Pred_{ij}$ | predicate | nil | predicate $j$ in module $i$ |

**invariants**

| | |
|---|---|
| $\mathcal{IMP}_i = \{Pred_{ij} \in MO_i \wedge import(Pred_{ij})\}$ | imported predicates in $MO_i$ |
| $\mathcal{EXP}_i = \{Pred_{ij} \in MO_i \wedge export(Pred_{ij})\}$ | exported predicates in $MO_i$ |
| $\mathcal{GLP} = \bigcup\limits_{1 \leq i \leq m} EXP_i$ | all exportable predicates |
| $\mathcal{IMP}_i \subseteq \mathcal{GLP}$ | imported predicates are global |

**computation**

| | |
|---|---|
| $\parallel \quad \langle \quad i : MO_i \in \mathcal{MO} ::$ | for all modules |
| $\qquad \parallel \quad P_i \quad := \quad \text{EVALMOD}_i(MO_i, \mathcal{IMP}_i)$ | execute module $MO_i$ using imported predicates $\mathcal{IMP}_i$ |
| $\quad \rangle$ | |

**end MODULE-EXEC1**

Figure 4.4: Program MODULE-EXEC1

*program* is defined as a set $\mathcal{MO} = \{MO_1, \ldots, MO_m\}$ of modules, together with a set of global predicates

$$\mathcal{GLP} = \bigcup\limits_{1 \leq i \leq m} \mathcal{EXP}_i$$

where $\mathcal{EXP}_i$ designates the **exportable** predicates of module $MO_i$, and $\mathcal{IMP}_i$ its **imported** predicates together with their owners (the modules where the predicate is defined as **exportable**).

This solution admittedly is not very satisfying; all it achieves is the accessibility of predicate definitions from outside. In principle, this is only a macro expansion facility which could as well be handled by copying the required predicates at compile time into the code of the corresponding module, and the exchange of actual parameter values is not possible.

### Correctness and Complexity

There are no problems concerning the correctness of programs composed of modules and evaluated according to the program specified in Figure 4.4; a modular program according to the paradigm used there is equivalent to a program where the imported predicates are copied explicitly into the program. The run-time complexity of a modular program is the same as with explicit copies of predicates, assuming that the action of importing predicates is negligible compared to the actual computation performed.

### Internal Parallelization

The possible internal parallelizations for the evaluation of modules depend strongly on the actual evaluation mechanism used; a number of candidates will be discussed in the later sections of this chapter.

### Refinements

Exchanging values is possible if not only the definition of a predicate is taken into account, but also its current parameter values, or variable substitutions; this is designated by $\sigma_i Pred_{ij}$, expressing that variables of $Pred_{ij}$ are instantiated according to the substitutions defined by $\sigma_i$. Here we encounter another problem, which is caused by the non-determinism potentially enclosed in the execution of logic programs:

- which definition of a predicate is used in the case that there is more than one;
- which substitution is chosen if OR-parallelism is in effect;
- what happens if back-tracking has to be performed, and substitutions have to be discarded?

The solution to this problem is to use a set of substitutions, corresponding to the alternative solutions for the module $MO_i$ in which $Pred_{ij}$ is defined as **exportable**. It has to be taken into account here that alternatives not only may occur through multiple definitions for $Pred_{ij}$, but also through multiple definitions of predicates (of the same module) using $Pred_{ij}$.

There are a number of ways to exchange the current values of the substitutions:

- data-driven
  - global data: there is a global "pool" of predicates which is updated when a substitution which affects an exportable predicate is changed in a module;
  - copies: an updating of a substitution is broadcasted to all modules affected (where the corresponding predicate is imported);
- demand-driven
  - global data: as soon as a request for a certain predicate comes, it is updated;
  - pointers: if a predicate is used in a module where it is imported, its current parameter values are requested from its origin (the module where it is exported).

In the program MODULE-EXEC2 the relations between the imported and exported predicates are defined via a set of equations: for each imported predicate, its origin is specified as the module from where it is exported; it is also always up to date with respect to the current parameter values (by applying the substitutions from its origin module). This program works in a data-driven way using copies of the predicates together with the current parameter values. Note that there are two important differences between MODULE-EXEC1 and MODULE-EXEC2:

- the first program uses only the *definitions* of predicates to be imported, whereas the second program also takes the actual parameters into account;
- the first program relies on a global set of predicates composed of all exportable ones, whereas in the second program the relations between importable and exportable predicates are directly expressed through a set of equations defined as invariants.

Especially the first item is very essential since the second program allows for the *synchronization* of modules through their actual parameters; in the first program, it is not possible to express synchronization properly.

Due to possible interactions at execution time, questions regarding the issues of correctness and complexity are more difficult to handle

**Program MODULE-EXEC2**
**parameters**

| name | type | initially | description |
|---|---|---|---|
| $\mathcal{MO}$ | set of modules | | modules to be executed |
| $MO_i$ | module (set of clauses) | | single module |
| $P_i$ | proof | | output data of $MO_i$ |
| EVALMOD$_i$ | program | | evaluation mechanism for $MO_i$ |

**pre-conditions**

$\mathcal{MO} \neq \{\}$ — there are some modules

EVALMOD$_i \neq \{\}$ — evaluation mechanism for $MO_i$ is not empty

**local variables**

| | | | |
|---|---|---|---|
| $Pred_{ij}$ | predicate | nil | predicate $j$ in module $i$ |
| $\sigma_i$ | variable bindings | nil | current substitution for $MO_i$ |

**invariants**

$\langle\ \ i,j,k,l : Pred_{ij} \in MO_i \wedge import(Pred_{ij}) ::$   for all imported predicates

$\qquad Pred_{ij} = \sigma_k(Pred_{kl}) \wedge i \neq k \wedge export(Pred_{kl})$  define the origin

$\rangle$

**computation**

$[\!]\ \ \langle\ \ i : MO_i \in \mathcal{MO} ::$   for all modules

$\qquad [\!]\ \ P_i\ \ :=\ \ \text{EVALMOD}_i(MO_i)$   execute module $MO_i$

$\rangle$

**end MODULE-EXEC2**

Figure 4.5: Program MODULE-EXEC2

in the refined approach. What is required here is are methods to reason about temporal properties of the execution of programs; UNITY provides a number of basic constructs for this purpose. At this place, however, no general statements can be given since both correctness and complexity depend on the evaluation mechanism used.

### Combinations

Apart from combining a number of modules in a flat structure, it surely makes sense to provide a hierarchical structure. A denotation for the composition of modules is either implicitly given through the program text (e.g. the definition of a module may contain a 'call' of another module), or explicitly by composition operators for modules, defining an algebra of modules. The composition of modules into complex net-

| MODULARITY | | |
|---|---|---|
| | **data structures** | **operations** |
| **compu-tation** | **input data**<br>• formula (set of modules)<br>• sets of parameters (proof depth, number of inferences, number of copies), possibly different for each module<br>**output data**<br>• proof result (subresults of the modules)<br>• proof information (trace substitutions, proof tree)<br>**evaluation support**<br>• interdependencies between modules (sequence, hierarchy, synchronization)<br>• administrative inform.: allocation of the modules, shared resources, system status information | **remote procedure call**<br>• non-local evaluation of modules<br>**evaluation mechanisms**<br>• local evaluation of modules<br>**evaluation support**<br>• module administration (program structure)<br>• module allocation<br>• resource administration |
| **repre-sen-tation** | **evaluation mechanisms**<br>• code for the single provers<br>**global data**<br>• imported / exported pred.<br>**administrative data**<br>available resources, topology, activities<br>**evaluation support**<br>• status information | **modules**<br>• access to the code of the modules<br>**administrative data**<br>• access to the data<br>**evaluation support**<br>• memory management (virtual memory)<br>• garbage collection |
| **com-muni-cation** | **implicit**<br>• shared data structures, e.g. imported / exported predicates<br>**explicit**<br>• parameters<br>• messages | **implicit**<br>• shared data access<br>**explicit**<br>• parameter passing<br>• message passing |

Table 4.2: Modularity

works can be handled reasonably well as long as the structure can be described statically; dynamically changing structures, however, may cause problems both from the semantical side and for implementation.

A combination of modules and multitasking apparently is possible by assigning a set (or network) of modules to a task for execution.

## Conclusions

The description of parallel systems with the means of logic can be based on modules for introducing possibly hierarchical and dynamical structures in logic programs. Simple concepts like flat modularization can be realized easily (sometimes even in the compilation process); their expressiveness, however, is also not very high. More powerful approaches featuring hierarchical or dynamical structuring in principle require higher order logic to express properties of predicates; for dynamical structuring, temporal information about the execution of programs must be integrated in addition. At this level, modularity in logic programs has a close resemblance to the integration of object-oriented paradigms in logic.

## 4.6  Precision

The evaluation of a logic program can be pursued on different levels of exactness, filtering out invalid ones stepwise by increasing the effort invested. Various methods can be used to achieve these different levels, some have already been investigated by (Plaisted, 1981; Plaisted, 1987), but without particular attention to parallelism. The main approach applied in this book is based on paying attention to more or less details in the unification process (Lassez et al., 1986; Knight, 1989). Different unification mechanisms are used in the process of unifying two literals, ranging from a mere comparison of the predicate symbols (which corresponds to a restriction to propositional logic), to full unification including occur check. Thus unification is performed with various restrictions, resulting in the following levels of accuracy:

- *full unification* for predicate logic, which is correct and complete;
- *weak unification* (pairwise, statical), which is incorrect, but complete;
- *unification without occur-check* ("matching"); incorrect and incomplete (cyclic structures may cause non-terminating unification);
- *directed unification*, where the direction of the substitution process is indicated by the user (e.g. via read-only variables);
- *term matching*, where only one of the terms to be unified may contain variables;
- *no unification*, resulting in a restriction to propositional logic.

The problem is that a higher degree of exactness can be much more expensive in terms of memory and execution time. Focusing on parallelism, a layered execution strategy looks quite promising: A propositional logic prover is used to check out the propositional structure of the formula; its results in turn are used to guide the proof process of a prover using a higher degree of exactness, which in the extreme case that the formula is not valid can lead to an abortion of the proof process. Statical analysis of the program can reveal restrictions in the formula to be treated so that full unification is not necessary; this can be supported by annotations in the program itself, e.g. read-only variables.

**Problem Specification**

Define a proof mechanism EVALPREC which evaluates a logic formula

$F$ with different degrees of exactness, using appropriate unification mechanisms $UNIF_i$.

## Design of a Solution

As a first approach a single proof mechanism EVALPREC is used, and various degrees of exactness are introduced via different unification mechanisms. No information is exchanged between the different proof approaches. A proof approach $PA_i = (F, EVALPREC, UNIF_i)$ thus evaluates the formula $F$ according to EVALPREC and $UNIF_i$.

**Program PRECISION**

**parameters**

| name | type | description |
|------|------|-------------|
| $F$ | formula (set of clauses) | single module |
| $\mathcal{PA}$ | set of proof approaches | proof approaches pursued |
| $PA_i$ | proof approach | single proof approach |
| EVALPREC | program | evaluation mechanism |
| $UNIF_i$ | program | unification mechanism |
| $P_i$ | proof | proof (result) |

**computation**

| | |
|---|---|
| ‖ 〈 $i : PA_i \in \mathcal{PA} ::$ | for all proof approaches |
| ‖ $P_i := EVALPREC(F, UNIF_i)$ | evaluate formula $F$ |
| 〉 | |

**end PRECISION**

Figure 4.6: Program PRECISION

## Correctness and Complexity

Assuming that the proof mechanism EVALPREC itself is correct, the correctness of the single proof approaches depends on the unification mechanism used. If correctness is required, full unification including occur check must be used eventually.

The overall range of complexity for the different proof approaches reflects one major difference between propositional and predicate logic: Since there is no general method to decide the validity of quantified expressions, first order predicate logic is basically undecidable; if an expression is valid, then a procedure exists for verifying the validity of that expression; if the expression is not valid, the procedure may never terminate. For propositional logic, however, such a method to decide the validity exists (e.g. truth tables), and it is *decidable*.

If unification is regarded in isolation, *linear* algorithms have been proposed for full unification (Martelli and Montanari, 1982; Paterson and Wegman, 1978); their initial overhead, however, is quite high so that for practical purposes their suitability is questionable (Corbin and Bidoit, 1983), and in addition correctness is often sacrificed in favor of efficiency.

## Internal Parallelization

The potential for parallelization within the single proof approaches again depends on the particular approach used, which will be discussed in the next section. The parallelism possible during unification is investigated in the sections dealing with parallelism on the term level towards the end of this chapter.

## Refinements

Finding a proof for a logic program can be seen as traversing a search space spanned according to the properties of the formula; usually there is only little, information available to guide this search in order to avoid 'dead' branches in favor of 'prolific' ones. Such an information can be provided using the different degrees of exactness described above. Consider two extreme cases, no unification (propositional logic) and full unification. The first prover, which does not have to deal with the costly task of unification at all, of course can be much faster than the second one performing full unification. Thus the first one can gain some information during its proof process which can be valuable for the second one; an example are alternative branches which do not lead to a solution. The problem occurring here is that the exchange of 'useful' information is by no means trivial to organize; a basic idea is to incrementally add more 'flesh' to a skeleton provided by a prover working on a lower level of exactness.

Another obvious improvement can be achieved by not only using different unification mechanisms, but also different proof mechanisms or calculi on the various levels of exactness. A proof mechanism, for example, which is quite well suited for a full first order logic prover not necessarily is appropriate for propositional logic. The use of different proof mechanisms to introduce parallelism will be treated in the next section, which deals with competing calculi working on the same formula.

## Combinations

A combination of different provers with various degrees of exactness is feasible in two ways; they could run side by side on one and the same

| PRECISION | | |
|---|---|---|
| | **data structures** | **operations** |
| **compu-<br>tation** | **input data**<br>• logical formula (program)<br>• evaluation method applied<br>• unification mechanism<br>**output data**<br>• proof result (truth value)<br>• exactness / reliability<br>• options (trace, proof tree, evaluation method used, substitutions)<br>**intermediate data**<br>• strategic information: 'dead' branches, 'prolific' branches, important parts of the program (cycles), (partial) substitutions<br>**evaluation support**<br>• administrative information | **provers** with different precision / evaluation methods<br>• propositional logic prover<br>• PROLOG<br>• PROLOG with occur check<br>• Horn clause logic prover<br>• full first order logic prover<br>**unification mechanisms**<br>• full unification<br>• weak unification<br>• unification without occur check<br>• directed unification<br>• term matching |
| **repre-<br>sen-<br>tation** | **provers**<br>• code of the single provers<br>**unification mechanisms**<br>• code for unification<br>**administrative data**<br>• addresses of cooperating provers,<br>• activities of other provers | **access to data**<br>**access to code** |
| **com-<br>muni-<br>cation** | **implicit:**<br>• shared data structures<br>**explicit:**<br>• messages | **implicit**<br>• shared data access<br>**explicit**<br>• message passing |

Table 4.3: Precision

formula, possibly exchanging some intermediate results[3], or they could form a team where less exact provers superficially check the terrain for more exact ones organized in a hierarchical structure.

**Precision - Modularity** A combination of precision and modularity can occur on two levels: A program to be proven can be mapped onto modules or tasks with different evaluation mechanisms incorporating different degrees of exactness, or a program already structured into modules can be evaluated by applying multiple proof mechanisms with different exactness to the single modules. These two alternatives can also occur simultaneously, leading to the creation of additional modules (to be evaluated with different exactness) for a program composed of modules.

**Precision - Multitasking** Basically similar to the combination with modularity, precision can be intertwined with multitasking; remember that according to our understanding of multitasking there is no exchange of information between the single tasks. This choice might be more favorable in the case of large or complicated programs where an interaction between provers with different exactness does not seem to be helpful. In this case, the relations between the different degrees of exactness have to be identified by the user; this could be interesting in a case where one wants to estimate roughly if it is worth while to invest major efforts into the detailed evaluation of a computationally expensive program.

### Conclusions

The main importance of the precision category lies in the area of program development and programming methodologies by providing a method to investigate important properties of a program with increasing degrees of detail and effort. In combination with parallel evaluation of logic programs, it offers a good way to gather information for directing the proof process, e.g. which branch to choose.

---

[3] this can also be viewed as an instance of *cooperation*

## 4.7  Competition

Although the competition category in general comprises a wide spectrum of possibilities, our concentration on logic and parallelism here results in a restriction to the discussion of two important instances of competition for the evaluation of logic programs, namely the use of various *calculi*, and the application of different *strategies* and *heuristics* within one calculus.

### Calculi

A logical formula or program can be evaluated in a number of ways, directed by the *calculus* used for the evaluation. After a long history of not fully satisfying efforts (some prominent ones being Aristoteles' *syllogistics* and Leibniz' *lingua characteristica, calculus ratiocinator*) the foundations for a formalization of logic have been laid towards the mid of the last century by the work of A. De Morgan (De Morgan, 1847), G. Boole (Boole, 1854) and especially G. Frege (Frege, 1879). Around the turn of the century G. Peano created a more comprehensible notation, which was also the basis for A. N. Whitehead and B. Russell's epochal *Principia Mathematica* (Whitehead and Russell, 1910).

Since it is not feasible to investigate all the different claculi in detail here, we will concentrate on relevant features of calculi, especially with respect to automatization and parallelization. Important *features* of logical calculi, characterized by pairs of opposite terms, are:

**strong *vs.* weak:** A calculus $CAL_s$ is (informally) defined to be *stronger* than a calculus $CAL_w$ if for each formula $F_i$ the minimal length of a proof found with $CAL_s$ is smaller or equal to the one found with $CAL_w$; (Cook and Reckhow, 1974) give some results on the length of proofs for propositional calculi. A similar relation between two calculi can be described based on *simulation*: in these terms, $CAL_s$ is stronger than $CAL_w$ if each proof found with $CAL_w$ can be simulated polynomially with $CAL_s$ (Eder, 1989). A strong calculus uses few, but powerful deduction steps, whereas a weak one uses simple deduction steps, but as a consequence needs more steps to find a proof.

**analytic *vs.* synthetic:** An *analytical* calculus leaves the formula to be proven unchanged during the proof process; a *synthetical* calculus manipulates the formula by modifying or deleting parts of it, or by creating new subformulae. Especially for automated theorem provers this differentiation is somewhat artificial since in the analytical case the information gained in the proof process is not stored within the formula itself, but in some additional data structures, and in an actual implementation it might be difficult to clearly separate these two.

**normal form** *vs.* **non-normal form:** Some calculi can only operate on formulae in some *normal form*[4], whereas others are able to treat formulae in arbitrary form.

**skolemized** *vs.* **non-skolemized:** *Skolemization* designates the process of eliminating existential quantifiers from a formula by replacing the variables in their scope with constants or functions (called Skolem functions). It is of special importance for the automation of calculi since it allows more efficient proof procedures.

Of special interest in our case are two more properties, namely the potential for *automation* and for *parallelization* inherent in a calculus. A calculus may be suited for automation if

1. both its deduction step(s) and a strategy to choose an appropriate step can be described with moderate effort in form of an algorithm, and

2. the execution of a single deduction step is sufficiently efficient, and

3. the strategy leads to reasonably short proofs.

A calculus is a good candidate for parallelization if the required data structures and the algorithmic structure of the calculus permit a clear separation of independent parts. In particular, an important criterion for parallelization of a calculus is the number of deduction steps which can be applied independently at a given point during the execution; this criterion, however, is difficult to capture since it depends on the formula to be evaluated as well. Restrictions on the use of parallelism can be given by the need of global data for the proof process (as in the resolution calculus, for example), or the occurrence of shared variables. These properties are, to a certain degree, related to the ones mentioned above; a strong calculus, for example, in general is less open

---

[4] familiar ones are *conjunctive* or *disjunctive* normal form, others are *prenex*, *clause* and *definitional* (Eder, 1985) form; the latter is of special interest since the length of the formula increases only linearly in the transformation process

to automation than a weak one, and its potential for parallelization is difficult to capture. For more information and a more formal treatment of some of the questions addressed here see (Eder, 1989; Bibel, 1990).

## Problem Specification

A logical formula $F$ is to be evaluated by simultaneously using different calculi $CAL_i$.

## Design of a Solution

A straightforward solution consists of independently applying a number of evaluation mechanisms EVALCAL$_i$, based on calculi $CAL_i$, to (copies of) the formula $F$. This can be seen as a special case of the *multitasking* paradigm, identifying a task as the formula together with an evaluation mechanism based on a calculus.

**Program CALCULI**

**parameters**

| name | type | description |
|------|------|-------------|
| $TA$ | set of tasks | tasks to be executed |
| $TA_i$ | task | single task |
| $F$ | formula (set of clauses) | only one formula |
| $P_i$ | proof | result (proof) |
| EVALCAL$_i$ | program (set of statements) | evaluation mechanism of $TA_i$ |

**pre-conditions**

| | |
|---|---|
| $TA \neq \{\}$ | there are some tasks |
| $F \neq \{\}$ | formula not empty |
| $P_i = \{\}$ | no proof |
| EVALCAL$_i \neq \{\}$ | evaluation mechanism not empty |

**invariants**

| | |
|---|---|
| $TA_i = (F, \text{EVALCAL}_i, P_i)$ | components of a task |

**computation**

```
∥  〈  i : TA_i ∈ TA ::              for all tasks
       ∥ P_i  :=  EVALCAL_i(F)       evaluate formula F
    〉
```

**post-conditions**

| | |
|---|---|
| $P_i \neq \{\}$ | a proof has been found for $F_i$ |

**end CALCULI**

Figure 4.7: Program CALCULI

### Refinements

The above solution can be improved in at least two directions:

- cooperation: some subresults (e.g. lemmata) can be shared once they have been found;
- analysis: before evaluation the formula can be analyzed to identify appropriate calculi.

### Combinations

The concurrent evaluation of a formula by provers based on different logical calculi is adequate if chances are good that the evaluation processes of the different provers might be of considerably differing complexity; in other words this means that a proof of a formula is quite easy with one calculus, whereas it is more difficult, or much longer, with a different one. In general, this is not easy to decide for a particular formula, and so it might be useful to use various calculi for proof approaches.

A basis for such an approach can be found in so-called *inference labs*, where a number of provers with different evaluation features are provided within a common environment (Systems Abstracts in (Lusk and Overbeek, 1988; Stickel, 1990a)); (Niemilae and Tuominen, 1987)). They are usually based on sequential execution, allowing only to check out various alternatives one after the other. In some cases, however, even a pseudo-parallel execution, e.g. via multitasking on conventional systems, can be of advantage: assume the extreme case in which a formula can be proven with one calculus in a single step, whereas with another calculus it requires (infinitely) many steps. The total execution time in the sequential case is the sum of the single execution times (which can be quite large, if an unfavorable sequence is chosen), whereas in the pseudoparallel case it is the shortest execution time times the number of different proof approaches.

**Calculi - Precision**  A first combination of calculi with different degrees of precision is obtained by applying the concepts described in the previous section. Basically this means to use different *unification* algorithms in a calculus. A second possibility is to obtain 'sloppy' versions of a calculus by modifying the *inference rules* towards better efficiency. In both cases occurs the danger that correctness is affected; it has to be guaranteed that there is at least one correct prover available, and all results of (potentially) incorrect provers have to be confirmed

| CALCULI | | |
| --- | --- | --- |
| | **data structures** | **operations** |
| **computation** | **input data**<br>• basic inference rules<br>• formula (program),<br>• calculus (inference mechanism, theory processor)<br>**output data**<br>• proof result (truth value)<br>• options (substitutions, proof tree, calculus used, trace, statistical and heuristic information)<br>**intermediate data**<br>• successful / failed approaches<br>• calculus-dependent information<br>• heuristic information<br>**evaluation support**<br>• administrative information | **provers**<br>• with different underlying calculi<br>**basic deduction steps** |
| **representation** | **provers**<br>• code of the single provers<br>**administrative data**<br>• addresses of cooperating provers<br>• other calculi used<br>• status of other provers<br>**evaluation support**<br>• heuristic information | **provers**<br>• access to the code of the provers<br>**data**<br>• access to the data |
| **communication** | **implicit**<br>• shared data structures<br>**explicit**<br>• messages | **implicit**<br>• shared data access<br>**explicit**<br>• message passing |

Table 4.4: Calculi

by a correct one or by additional correctness checks.

**Calculi - Modularity**   Using metaprogramming techniques or
theories provides a possibility to formulate specific inference mechanisms, which can be based on appropriate calculi, for each module of
a complex program. This is of interest for tasks where modules with
specific features can be identified, such as in knowledge processing
systems for specialized applications. The other way round, specific
inference mechanisms (theory processors) can be offered as services of
a logic programming system in form of predefined library modules.

**Calculi - Multitasking**   Provided that there is no requirement for
information exchange between different proof approaches, the concepts
described in the above paragraph for modularity can be realized via
multitasking in a similar way.


## Strategies

The search space to be traversed during the evaluation of a logic formula can be extremely huge, and a brute force method checking all
possibilities rarely can be applied. In some cases, properties of both
the problem to be solved (here represented by the logic formula) and
the method used to solve the problem (here given by the calculus),
can be taken into account to guide the search process (Trapp, 1987).
Such a guidance can be expressed through strategies and heuristics;
a general description of problem solving in Artificial Intelligence via
heuristic search can be found in (Kaindl, 1989).

A *strategy* is a global directive for the search process, aiming at
an optimal or satisfying effort by analyzing properties concerning the
search process for a given formula as a whole. A strategy is defined by
a rule which is applied and evaluated in each step of the proof process.
Typical instances for search strategies are depth / breadth first and
top down / bottom up (for trees).

*Heuristics* typically are used to make decisions about the next step
to be taken by an analysis of the local environment, i.e. information
on previous steps together with features of the part of the formula
which is currently under evaluation. A solution to the problem is obtained by reasoning from past experience since no algorithm is known
or is relevant. Heuristics are defined by a function which chooses the
best out of a number of alternatives; for this purpose, an indicator

for the quality of a solution according to a certain criterion is computed, typically making use of information gathered from previous attempts, data obtained so far in the present attempt, and features of the alternatives under consideration. Applications for heuristics in the evaluation of logic formulæ are the support of a decision in the choice of OR-branches, or in a rearrangement of rules and subgoals.

With respect to parallelism, strategies offer an opportunity to derive a set of competing provers from one and the same basic calculus by applying different strategies within that calculus. With heuristics the case is a little different; the goal here is to choose the best one out of a number of viable alternatives. Thus, they preferably can be used to restrict the exploitation of parallelism to a reasonable degree, but they also can be used to introduce parallelism via competing provers (a strategy can consist of applying heuristics).

A generic operation for the evaluation of a logic formula is to search a graph (usually a tree) for paths with a certain property. As a first approach we describe a general graph search program, which will be refined in later steps. It is based on a number of concepts briefly described beforehand:

**Definition 4.1 (Graph)**

A graph $G = (\mathcal{N}, \mathcal{E})$ consists of a nonempty set $\mathcal{N}$ of *nodes* and a set $\mathcal{E}$ of edges. An edge $E_{ij}$ is defined as a pair of nodes $(N_i, N_j)$. In a *directed graph* the edges are represented as ordered pairs of nodes with a relation $\prec$ defining the direction between the nodes. For an edge $E_{ij}$ with $N_i \prec N_j$ the node $N_i$ is called *predecessor* of $N_j$, and $N_j$ *successor* of $N_i$. In addition to their names and indices, both nodes and edges may contain additional information (e.g. the length of an edge). ∎

**Definition 4.2 (Path)**

A path from node $N_i$ to $N_l$ is a sequence of edges such that for each successive pair of edges $(E_{ij}, E_{kl})$ in the sequence $j = k$; a path can also be described by the sequence of nodes it consists of. ∎

**Definition 4.3 (Cycle)**

A cycle is a path starting and ending at the same node. ∎

**Definition 4.4 (Tree)**

A tree is a (connected) graph without cycles. Usually a tree is defined as a set of nodes $\mathcal{N}$, a distinguished element $r \in \mathcal{N}$, the root of the tree, a distinguished subset $\mathcal{L} \subset \mathcal{N}$, and a relation $\prec$ defining the (directed) edges between the nodes. A path $N_i, \ldots, N_k$ in a tree is defined through the relation $N_j \prec N_{j+1}$ for all $i \leq j \leq k$; if $N_i \equiv r$ and $N_k \in \mathcal{L}$ the path is said to be a *branch* of the tree.             ∎

## Definition 4.5 (Connectivity Matrix)

Especially for computation purposes a graph with $n$ nodes can be represented as a $n \times n$ matrix $E$ where $E[i,j] \in \{\text{true}, \text{false}\}$; instead of simple boolean values the elements of the matrix may contain as well other information like the length of the shortest path between node $N_i$ and $N_j$.                                                      ∎

## Problem Specification

Evaluate a formula $F$ by concurrently feeding it to proof mechanisms based on one and the same calculus, differing in their strategies to traverse the search space, and / or the use of heuristics.

## Design of a Solution

The central idea in the solution is to introduce a partial ordering on the graph to be searched in order to be able to choose one (or a subset) from a set of nodes to be treated next. Such a partial ordering can be determined statically, (e.g. for depth-first or breadth first search), or dynamically during the evaluation (e.g. using heuristical information). A graph containing cycles has to be mapped onto a directed acyclic graph to be able to introduce a partial ordering. The different strategies to be exploited are based on different partial orderings. Note that in this section we are interested in searching a graph concurrently with different search paradigms (strategies); we do not explicitly concentrate on parallel graph search techniques (for an overview of these see (Kronsjo, 1985)).

The central idea of the program STRATEG (Figure 4.8) is to describe a generic mechanism for searching a graph and computing certain properties of the paths it comprises. The search starts form a set $\mathcal{SOUR} \in \mathcal{N}$ of source nodes, proceeding along the (outgoing) edges of these nodes until the $\mathcal{DEST} \in \mathcal{N}$ of destination nodes is reached. A *search strategy* different from the brute force method of evaluating all possible path extensions simultaneously can be defined by selecting one (or a subset) of nodes according to a certain criterion. As examples, two programs incorporating breadth-first and depth-first strategies will be derived from the generic one.

**Program STRATEG**

**parameters**

| name | type | description |
|---|---|---|
| $\mathcal{G}$ | graph | graph to be traversed |
| $\mathcal{N}$ | set of nodes | nodes of the graph |
| $E[N, N]$ | connectivity matrix | matrix with the edges |
| $\mathcal{SOUR}$ | source nodes | start nodes |
| $\mathcal{DEST}$ | destination nodes | end nodes |
| $P[N, N]$ | property matrix | matrix with properties of paths |

**pre-conditions**

$\forall i, j :: P[i, j] = \mathsf{false}$     no paths checked yet

**local variables**

$stat$ array $[N]$, $stat[i] \in \{$source, dest,     status of the nodes
settled, pending, rest$\}$

$\quad stat[i] = \mathsf{source}$ if $i \in \mathcal{SOUR}$

$\quad stat[i] = \mathsf{dest}$ if $i \in \mathcal{DEST}$

$\quad stat[i] = \mathsf{rest}$ if $i \in \mathcal{N} \setminus (\mathcal{SOUR} \cup \mathcal{DEST})$

**invariants**

$\forall i, j : P[i, j] = \mathsf{true} :: \exists k : stat[k] = \mathsf{source}$     all nodes in a path

$\quad \wedge P[i, k] = \mathsf{true} \wedge P[j, k] = \mathsf{true}$     must be reachable

$\forall i : stat[i] = \mathsf{pending} :: \exists j : P[i, j] = \mathsf{true}$     pending node in a path

$\forall i : stat[i] = \mathsf{settled} :: \exists j : P[i, j] = \mathsf{true}$     settled node in a path

$\forall i, j : stat[i] = \mathsf{rest} :: P[i, j] = \mathsf{false}$     rest node not in a path

**computation**

$\| \ \langle \quad i, j, k : \ stat[i] = \mathsf{pending} \wedge E[i, j] = \mathsf{true} ::$     all nodes $i$ on an open end of a path, $i$ connected to $j$

$\quad \| \ stat[i] := \mathsf{settled}$     $i$ is inside a path now

$\quad \| \ stat[j] := \mathsf{pending}$ if $stat[j] = \mathsf{rest}$     $j$ is a new open end

$\quad \| \ P[k, j] := P[k, i]$     if there was a path from $k$ to $i$, then there is also a path from $k$ to $j$

$\rangle$

**post-conditions**

$\forall i, j : stat[i] = \mathsf{source} \wedge P[i, j] = \mathsf{true} ::$     all nodes

$\quad stat[j] = \mathsf{settled}$     have been visited

**end STRATEG**

Figure 4.8: Program STRATEG

Let the nodes $\mathcal{N}$ in the graph be numbered $0 \ldots N - 1$; for reasons of convenience, an additional node $N$ is introduced which never is included in any path, but has an edge from each other node. The edges are described in a connectivity matrix[5] $E[N, N]$, with $E[i, j]$ is **true** if there is an edge connecting node $i$ and node $j$. A similar matrix can be used to store the properties of the paths to be evaluated; here we use a boolean matrix $P[N, N]$ which simply indicates that a path from a node $k$ to a node $l$ already has been checked by assigning the value **true** to the corresponding element $P[k, l]$. Instead of explicitly using sets to describe certain features of the nodes, a status array $stat[0 \ldots N]$ is used with the following properties of nodes:

- **source**: This node represents a starting point for the search.

- **destination**: End point of the search.

- **settled**: A node is said to be **settled** if all its edges have been examined.

- **pending**: A node with some examined edges, and some edges not treated yet is called **pending**; this corresponds to an open path, which can be extended via these edges.

- **rest**: All nodes not reached yet.

It is possible to encode the **source** and **destination** properties using **settled** and **pending**, depending if the graph is directed or not; for the sake of clarity, however, we use the explicit designation.

The goal is to have all nodes settled; a step towards it consists of examining all open edges of all **pending** nodes, changing the states of these nodes into **settled**, and determining a new set of **pending** nodes by following the open edges of the previous set of **pending** nodes.

## Correctness and Complexity

This program reaches (for finite graphs) a fixed point; with every state change, the **rest** is diminished by the new pending nodes, and the old pending nodes are added to **settled**. Once a fixed point has been reached, the following conditions hold:

- no pending nodes with open edges are left;

- all destination nodes which have a connection to a path are reachable from a source node.

---

[5] for sparsely connected graphs and for trees, a matrix of course is not the most memory-efficient way to store this information

The complexity of the program is $\mathcal{O}(N^3)$, $N$ being the number of nodes (including the dummy node introduced for convenience); the computations dealing with invariants are not included here.

## Refinements

In many cases it is useful to have a program which can deal with (limited parts of infinite) graphs without running forever; below two refined versions are given which deal with a limited number of successors to a set of nodes, and paths of limited length. Derived from the general program above, two specializations are regarded; one is restricted breadth-first, exploiting only a limited number of paths in parallel, the other bounded depth-first, limiting the depth of the graph (or length of a path) to be traversed.

**Restricted Breadth-first**   The breadth-first strategy is restricted to the extension of $p$ paths in parallel, where $p$ may be inspired from the fact that in reality only a limited number of processing elements are available. If graphs containing nodes with a potentially infinite number of edges are regarded as well, a restriction for breadth-first is necessary to maintain completeness of the search strategy.

In the program, a restricted breadth-first strategy can be expressed by introducing an additional property **selected** for the nodes, meaning that a subset of the **pending** nodes will be extended. The condition which of the **pending** nodes is selected can be given in an invariant and defines the actual strategy. Typically a partial ordering is imposed on the nodes, and the selection condition is formulated based on this partial ordering. In the following we use an integer parameter *breadth* to select a number of *breadth* nodes with the lowest values with respect to some partial ordering.

**Bounded Depth-first**   If a graph may contain paths of infinite length, completeness in the search process can be maintained by introducing a depth boundary which is increased after all paths of that length have been checked. A bounded depth first strategy is imposed onto the STRATEG program by always selecting some nodes with the highest depth value, but not exceeding the given boundary, from the pending nodes. Note that if only one node is chosen in each step, the evaluation is done in a sequential way; this case is specified in the program STRATBDF (Figure 4.10). The search strategy again is expressed in form of an invariant.

**Program** STRATRBF
**parameters**

| *name* | *type* | *description* |
|---|---|---|
| $\mathcal{G}$ | graph | graph to be traversed |
| $\mathcal{N}$ | set of nodes | nodes of the graph |
| $E[N, N]$ | connectivity matrix | matrix with the edges |
| $\mathcal{SOUR}$ | set of nodes | source nodes |
| $\mathcal{DEST}$ | set of nodes | destination nodes |
| $P[N, N]$ | property matrix | properties of paths |
| $breadth$ | integer | breadth restriction |

**pre-conditions**

$breadth > 0$        selection is possible

**local variables**

$stat$ array $[N]$, $stat[i] \in \{$selected, dest,      status of the nodes
           settled, pending, rest$\}$
     $stat[i] = $ selected if $i \in \mathcal{SOUR}$
     $stat[i] = $ dest if $i \in \mathcal{DEST}$
     $stat[i] = $ rest if $i \in \mathcal{N} \setminus (\mathcal{SOUR} \cup \mathcal{DEST})$

**invariants**

$| \{k : stat[k] = $ selected$\} | \leq breadth$      less selected nodes than
                                              $breadth$

$[\!] \ \langle \ \ i : stat[i] = $ pending $:: stat[i] = $ selected      from the pending nodes
     if $\forall j : j \neq i \wedge stat[i] = $ pending $::$      are selected
     $breadth[i] \leq breadth[j] \wedge$      the first
     $| \{k : stat[k] = $ selected$\} | \leq breadth$      $breadth$ nodes
   $\rangle$

**computation**

$[\!] \ \langle \ \ i, j, k : \ stat[i] = $ selected $\wedge E[i, j] = $ true $::$      all nodes $i$ on an open
                                                  end, $i$ connected to $j$

   $\|$ $stat[i] := $ settled      $i$ is inside a path now
   $\|$ $stat[j] := $ pending if $stat[j] = $ rest      $j$ is new open end
   $\|$ $P[k, j] := P[k, i]$      if there was a path from
                                               $k$ to $i$, then there is also
                                               a path from $k$ to $j$

   $\rangle$
**end** STRATRBF

Figure 4.9: Program STRATRBF

**Program** STRATBDF
**parameters**

| name | type | description |
|---|---|---|
| $\mathcal{G}$ | graph | graph to be traversed |
| $\mathcal{N}$ | set of nodes | nodes of the graph |
| $E[N, N]$ | connectivity matrix | matrix with the edges |
| $\mathcal{SOUR}$ | set of nodes | source nodes |
| $\mathcal{DEST}$ | set of nodes | destination nodes |
| $P[N, N]$ | property matrix | properties of paths |
| $breadth$ | integer | breadth restriction |
| $depthbound$ | integer | depth boundary |

**local variables**

$stat$ array $[N]$, $stat[i] \in \{$selected, dest,      status of the nodes
                  settled, pending, rest$\}$

     $stat[i] =$ selected if $i \in \mathcal{SOUR}$
     $stat[i] =$ dest if $i \in \mathcal{DEST}$
     $stat[i] =$ rest if $i \in \mathcal{N} \setminus (\mathcal{SOUR} \cup \mathcal{DEST})$

**invariants**

$| \{k : stat[k] =$ selected$\} | \leq breadth$      the number of selected node is smaller or equal to $breadth$

$\| \langle$   $i : stat[i] =$ pending $:: stat[i] =$ selected    from the pending nodes
       if $\forall j : j \neq i \wedge stat[i] =$ pending $::$    select the node with the
       $depth[i] \leq depth[j] \wedge$    highest $depth$ value, but
       $depth \leq depthbound$    smaller than $depthbound$
   $\rangle$

**computation**

$\| \langle$   $i, j, k : stat[i] =$ selected $\wedge E[i,j] =$ true $::$    all nodes $i$ on an open end, $i$ connected to $j$

     $\| \; stat[i] :=$ settled    $i$ is inside a path now
     $\| \; stat[j] :=$ pending if $stat[j] =$ rest    $j$ is new open end
     $\| \; P[k,j] := P[k,i]$    if there was a path from $k$ to $i$, then there is also a path from $k$ to $j$

   $\rangle$

**end** STRATBDF

Figure 4.10: Program STRATBDF

**Correctness and Complexity**

The modifications introduced in the two refined versions do not affect
the correctness of the programs. They enforce completeness of the
search for graphs of infinite breadth (a node has an infinite number
of direct successor nodes) or infinite depth (the length of a path is
infinite); they also can be combined into one single program to adapt
a search strategy to the number of processing elements in a given
system. The complexity of the refined programs is only different with
respect to the treatment of graphs with an infinite number of edges,
or with infinite paths (cycles): whereas the original version cannot
handle these cases, each of the refined versions can handle one.

**Combinations**

A problem whose structure is difficult to capture can be tackled by
provers incorporating different instances of the same strategy; an example for that has been derived from *iterative deepening*, where a team
of identical provers with a depth-bound strategy is used, but each with
a different value for its depth boundary. It is also possible to apply
different strategies in parallel, especially if strategies with diametral
features (like depth-first and breadth-first) are used together. A more
sophisticated approach is to use 'intelligent' strategies which make use
of information gathered during the proof process, and may adapt itself
dynamically to the structure of the computation; in our terminology,
this would be a combination of heuristics and strategies.


**Strategies - Calculi**  An obvious combination of strategies and
calculi is to use different strategies to refine a calculus into different
versions, if the basic specification of the calculus allows such modifications. Another possibility is to combine teams of calculi and strategies
in such a way that each pair can treat one particular class of problems
very efficiently. In total, such a combination yields a good mutual supplement: the calculus to define the steps from one node in the search
space to the next (computation rule), the strategy to choose one from
a set of possible successor nodes, and heuristics to improve this choice
by adapting it to the current situation (evaluation function).


**Strategies - Precision**  Strategies and precision can be combined
by using lower precision to get a good evaluation function for strategies
and heuristics with higher precision.

| STRATEGIES | | |
|---|---|---|
| | **data structures** | **operations** |
| **compu-tation** | **input data**<br>• formula (program),<br>• options (strategies to be applied, heuristics)<br>**output data**<br>• proof result (truth value)<br>• options (substitutions, proof tree, applied strategies, trace, statistical and heuristic information)<br>**intermediate data**<br>• heuristic information (promising strategies, 'dead' / 'prolific' branches)<br>• lemmata<br>**evaluation support**<br>• administrative information | **provers**<br>• with different evaluation strategies<br>**evaluation functions**<br>• for heuristic choices |
| **repre-sen-tation** | **provers**<br>• code of the single provers,<br>• specification of the strategies to be applied<br>• evaluation functions for heuristics<br>**administrative data**<br>• addresses of connected provers<br>• other applied strategies<br>• status of other provers<br>**evaluation support**<br>• strategy libraries<br>• predefined heuristics | **provers**<br>• access to the code of the provers, strategies and heuristics<br>**data**<br>• access to the data (`read/write`) |
| **com-muni-cation** | **implicit**<br>• shared data structures<br>**explicit**<br>• messages | **implicit**<br>• shared data access<br>**explicit**<br>• message passing |

Table 4.5: Strategies

**Strategies - Modularity**  A complex problem with an overall irregular computational structure can be decomposed into modules whose computational structure is comprehensible, and strategies can be mapped specifically onto those modules. Modules also provide an appropriate structuring mechanism if a problem is to be tackled by a team of interacting provers with different strategies. In more practical terms, module libraries with predefined strategies and heuristics can be used as a program development support.

**Strategies - Multitasking**  The combination of strategies and multitasking in general is similar to modules; the main difference is that in our definition of multitasking there is no interaction between single tasks (on the user level). In cases where the exchange of information is considered unnecessary, this combination can be adequate.

## Conclusions

The competition category is very promising for the parallel execution of programs. It provides the potential to deal with different classes of problems in an adequate way (i.e. by the choice of a suitable calculus), and enlarges the probability of finding the right path (the shortest proof) by the application of different strategies. In the PARTHEO system, a strategy based on bounded depth (iterative deepening) has been integrated with very good results (Schumann and Letz, 1990; Bayerl et al., 1989). The speed-up achieved by applying several different strategies in parallel can be extremely high, but depends heavily on the structure of the problem.

# 4.8   Spanning Sets

The identification of possible alternative solutions for the proof of a formula is examined in this section. It is based on an analysis of the formula which can be done largely at compile time. In this analysis, important information contained in the structure of the formula is gathered through *spanning sets*. In terms of the connection method, a set of connections $SPASE$ in a matrix $F$ is called spanning for $F$ if each path through $F$ contains a connection. A formula is valid iff there exists a spanning set of connections for it and a substitution such that each connection of the spanning set after the application of the substitution consists of a complementary pair of literals (Bibel, 1987). If more than one spanning sets – together with a corresponding substitution – can be found, each of them represents an alternative proof for the formula. The set of all spanning sets of a formula is designated $\mathcal{SPASE}$.

Parallelism can be introduced in two ways here; one is to identify candidates for spanning sets in parallel, the other is to find the substitutions necessary for a proof of the formula. In many approaches these two subtasks are not treated separately due to efficiency reasons, but for the sake of clarity they are split up in this discussion; proposals based on such a separation are also described in (Bibel et al., 1987; Balcke, 1987; Jorrand, 1989; Ibañez, 1989; Wang, 1989). These subtasks again can internally be treated in parallel.

### Problem Specification

As indicated above, the problem is divided into two subproblems which can be treated one after the other, and each for itself can be split up into independent instances of a task.

1. Identification of spanning set candidates:
   For a logical formula $F$ identify the set $\mathcal{CANSPS}$ of candidates for spanning sets.

2. Determination of substitutions:
   For each spanning set candidate, try to find a substitution $\sigma$ such that $F$ is valid. This is the task of unifying all the terms appearing in the connections of the spanning set candidate.

We will concentrate on the first part of the problem, since unification will be treated in the section on term parallelism.

### Design of a Solution

The process of identifying candidates for spanning sets of connections can be pursued in a way that substitutions are not dealt with[6], and then is independent of the term structure contained in the literals of the connections. Neglecting the term structure of a first order predicate logic formula corresponds to a restriction to propositional logic, where the literals are simply positive or negative propositional variables.

In principle, this task can be seen as searching a graph for paths with certain properties, or, more precisely, checking out if all paths through the graph derived from the formula have the property to contain a complementary pair of literals. Due to certain features specific to such a connection graph, a general graph search procedure may be easily outpowered by specialized ones.

The basic idea for the identification of candidate spanning sets is to incrementally add appropriate connections to an already generated subset. The problem lies in the choice of these appropriate connections to be added: cyclic chains of connections – or *cycles*, for short, – may occur (representing recursively structured formulae) and lead to infinitely many spanning sets for a formula. The problem with cycles, however, can be postponed by identifying *minimal* spanning sets (which must not contain cycles) and potential cycles occurring in this spanning set separately; it then has to be resolved during unification by the creation of clause copies.

The program SPASE (Figure 4.11) captures only the very essential framework for the spanning set approach; it does not apply a strategy to choose appropriate connections to spanning set candidates, nor does it avoid redundant work. In addition, it may never terminate since no check for cycles is performed.

### Refinements

An extensive treatment of an approach using spanning sets and the connection method has been pursued by M.-B. Ibáñez at LIFIA, Grenoble (Ibañez, 1989), and presents the basis for the information shown in Tables 4.6 - 4.8[7]; more general aspects regarding the use of spanning sets are resumed in Table 4.9.

---

[6] but it is also possible to use weak unification, for example

[7] to a certain degree the description there has been influenced by the use of the language FP2 as specification tool

**Program SPASE**
**parameters**

| *name* | *type* | *initially* | *description* |
|---|---|---|---|
| $F$ | set of clauses | | formula (as matrix) |
| *result* | {true, false} | | result of the evaluation |

**pre-conditions**

$F = \{CLAUSE_1, \ldots, CLAUSE_n\}$      clause notation

**local variables**

| $\mathcal{PATHS}$ | set of paths | { } | paths in the matrix |
|---|---|---|---|
| $PATH_i$ | set of literals | { } | single path |
| $\mathcal{LITERALS}$ | set of literals | { } | literals of the matrix |
| $L_l$ | $L_l = P_l(t_l)$ | nil | literal: predicate, term |
| $P_k$ | predicate | nil | pred. $k$ in the formula |
| $\sigma$ | variable bindings | nil | substitution |
| $\mathcal{CONN}$ | set of pairs of literals | { } | connect. in the formula |
| $CONN_i$ | pair of literals | (nil, nil) | single connection |
| $\mathcal{CANSPS}$ | set of sets of connect. | { } | spanning set candidates |
| $CANSPS_i$ | set of connections | { } | single candidate |

**invariants**

$$PATH_i \equiv \bigcup_{1 \leq j \leq |\mathcal{CLAUSES}|} L_j : L_j \in CLAUSE_j$$

definition path: one literal from each clause

$$\forall i,j : PATH_i, PATH_j \in \mathcal{PATHS} ::$$
$$PATH_i \neq PATH_j$$

all paths
are different

$$CONN_i \equiv (L_i, L_i') : L_i, L_i' \in \mathcal{LITERALS} ::$$
$$L_i = \neg L_i'$$

definition:
   connection

$$spanning(\mathcal{CONN}) \equiv \forall i : PATH_i \in F \; \exists j :$$
$$CONN_j \in \mathcal{CONN} :: CONN_j \in PATH_j$$

definition:
   spanning

$$unifiable(\mathcal{CONN}) \equiv \exists \sigma : \forall i :$$
$$CONN_i \in \mathcal{CONN} :: \sigma(t_i) = \sigma(t_i')$$

definition:
   unifiable

$$result = \bigvee_{i \in \mathcal{CANSPS}} result_i$$

disjunction of partial
results

**computation**

$$\| \langle \; i,j : CANSPS_i \in \mathcal{CANSPS}$$
$$\wedge CONN_i \in \mathcal{CONN} ::$$

for all candidate sets
and connections

$$\| CANSPS_i := CANSPS_i \cup CONN_j$$
$$\| result_i \;\; := spanning(CANSPS_i)$$
$$\wedge unifiable(CANSPS_i)$$

add connection to set
spanning and unifiable
set of connections

$$\rangle$$

**end SPASE**

Figure 4.11: Program SPASE

| FIND CONNECTIONS | | |
|---|---|---|
| | data structures | operations |
| **compu-tation** | **input data**<br>• formula<br>**output data**<br>• set of connections / (extended) connection graph<br>**intermediate data**<br>• set of clauses<br>• set of literals<br>• set of predicates | **connections**<br>• identify complementary pairs of literals<br>**structure**<br>• weak unification |
| **repre-sen-tation** | **program**<br>• code for the search for complementary pairs of literals<br>• code for the weak unification<br>**intermediate data**<br>• set of clauses<br>• set of literals<br>• set of predicates | **access to code**<br>**access to data** |
| **com-muni-cation** | **explicit**<br>• messages | **explicit**<br>• message passing mechanisms ($n$-party rendezvous) |

Table 4.6: Find Connections

Closely related work has also been described in (Balcke, 1987) and (Wang, 1989). As a common feature for these approaches, the computation of a proof for a given formula is done in the three phases

1. *find connections*: the connections which occur in the formula are identified; this is a simple search for literals with complementary predi cate symbols;

2. *construct spanning sets*: using the connections computed before, (candidates for) spanning sets are constructed; clause copies are made if their necessity can be determined here, otherwise connec-

tions potentially resulting in copies are stored separately for later treatment (cycles);

3. *unification*: for each resulting (candidate) spanning set, the most general unifier is computed; if it exists, the corresponding spanning set represents a proof for the formula.

| CONSTRUCTION OF SPANNING SETS | | |
|---|---|---|
| | **data structures** | **operations** |
| **compu-tation** | **input data**<br>• formula<br>• set of connections / (extended) connection graph<br>**output data**<br>• spanning sets of connections<br>**intermediate data**<br>• paths<br>• candidate sets | **construction**<br>• construct spanning set candidates by incrementally adding connections<br>**spanning**<br>• check if every path that goes through a clause has a connection in the spanning set<br>**cycles**<br>• check for cycles in the spanning set |
| **repre-sen-tation** | **program**<br>• code for the incremental construction of spanning sets<br>• code for the check of the paths<br>**intermediate data**<br>• set of paths | **access to code**<br>**access to data** |
| **com-muni-cation** | **explicit**<br>• messages | **explicit**<br>• message passing mechanisms ($n$-party rendezvous) |

Table 4.7: Construction of Spanning Sets

## Internal Parallelization

Parallelism can be detected in all three phases: In the first phase, each literal of the formula can be checked for complementarity against all others; it is questionable, however, if it is worthwhile to be exploited

| UNIFICATION | | |
|---|---|---|
| | **data structures** | **operations** |
| **compu-tation** | **input data**<br>• spanning set candidates<br>• logical formula (program)<br>• (extended) connection graph<br>• potential cycles<br>**output data**<br>• proof result<br>• substitutions<br>• clause / connection copies required<br>**intermediate data**<br>• term structure (*dag*) | **unification:** attempt to unify the terms of one specific spanning set candidate<br>• computation of variable bindings<br>• function evaluation<br>• occur check |
| **repre-sen-tation** | **unification mechanism**<br>• code for unification<br>**intermediate data**<br>• term structure (*dag*)<br>• subterms to be unified<br>• occur check status | **access to code**<br>**access to data**<br>• read *dag*<br>• modify *dag* |
| **com-muni-cation** | **explicit**<br>• messages | **explicit**<br>• message passing mechanisms (*n*-party rendezvous) |

Table 4.8: Unification

since the single tasks are very simple, and global information (the
whole formula) is necessary for each task. In the second phase, each
connection can serve as germ for a spanning set candidate, and the
further expansion of these candidates can proceed in parallel. This
endangers high redundancy however, since a spanning set candidate
typically comprises more than one connection, and the number of (different) candidates usually is much smaller than the number of connections. There is a trade-off between performing redundant work,
or detecting that redundant work occurs which requires information
about the overall work going on. A reasonable compromise is to re-

strict the amount of parallelism by a simple analysis of the formula. In our case, for example, distinct spanning sets can only crystalize from literals involved in more than one connection[8]. In the third phase, the obvious form of parallelism is to treat the different candidate spanning sets independently; in addition, the unification process itself can be parallelized.

## Combinations

The relevant parts of a logical formula for the construction of spanning sets are the connections of the respective formula, and a spanning set comprises a subset of the connections. Thus in principle the simultaneous construction of spanning sets can be initiated for each connection, resulting in a lot of redundant work, however. Filtering out these redundancies on the other hand necessitates a central instance with information about all ongoing work, which may turn out to be a bottleneck. The remedy can consist of a quick statical analysis of the formula under investigation for relatively independent parts and good initial crystalization points for spanning sets.

A nested hierarchy of spanning set constructors is only appropriate for calculi which allow the hierarchical composition of logical formulae (where a literal stands for a whole sub-formula).

**Spanning Sets - Strategies**  The construction of a spanning set can be viewed as proving a formula in two phases; first traversing the search space looking for solution *candidates*, then checking the correctness of such a candidate by unification. Thus a number of different strategies can be applied during the construction of a spanning set; a typical one is to split the connections involved into two subsets, one representing a *statical* part (where no cycles appear), the other a *dynamical* part potentially involving cycles and the creation of clause copies. For the second part, the creation of copies may depend on the results of unification, and the separation between spanning set construction and unification must be abandoned to a certain degree.

**Spanning Sets - Calculi**  The construction of spanning sets can be seen as a typical feature of (some) *analytical* calculi; it can also be used as an additional criterion in synthetic calculi while choosing favorable paths in the search space.

---

[8] this is also the source for OR-parallelism

| SPANNING SETS | | |
|---|---|---|
| | **data structures** | **operations** |
| **compu-tation** | **input data**<br>• formula<br>• options (evaluation method, precision, strategies, depth limit, number of solutions)<br>**output data**<br>• proof result (truth value)<br>• corresponding spanning set<br>• substitutions<br>• options (proof tree, evaluation method used, trace)<br>**intermediate data**<br>• spanning set candidates<br>• potential cycles<br>• connection graph<br>**evaluation support**<br>• administrative information | **identify spanning sets**<br>• connections<br>• *spanning* property<br>• potential cycles<br>**determine substitutions** (unification)<br>• computation of variable bindings<br>• function evaluation<br>• occur check |
| **repre-sen-tation** | **evaluation mechanism**<br>• program code<br>**data**<br>• formula<br>**intermediate data**<br>• spanning set candidates<br>• potential cycles<br>• connection graph<br>**administrative data**<br>• redundant work<br>• addresses of cooperating provers<br>• activities of other provers | **access to code**<br>**access to data** |
| **com-muni-cation** | **implicit**<br>• shared data structures<br>**explicit**<br>• copies of data structures<br>• messages | **implicit**<br>• shared data access<br>**explicit**<br>• data copying<br>• message passing |

Table 4.9: Spanning Sets

**Spanning Sets - Precision**   The selection of valid solutions from the candidates delivered by the spanning set construction process can be performed in a pipelined fashion, using 'sieves' with increasing precision[9]: a rough one filters out candidates with obnoxious flaws, e.g. incompatible term structures (weak unification); the next one eliminates candidates with occur check problems, and finally incompatible variable bindings are detected.

**Spanning Sets - Modularity**   Different spanning sets of a formula can be mapped onto modules or tasks, and then be evaluated in parallel. For a formula consisting of several modules, spanning sets can be constructed largely independently.

**Spanning Sets - Multitasking**   Instead of mapping spanning sets onto modules, they may as well be mapped onto different tasks, provided that no information exchange is envisaged between different tasks.

## Conclusions

The spanning set approach has a number of interesting features, especially in comparison with the typical PROLOG evaluation mechanism. It makes extensive use of a statical analysis of the formula to identify the spanning sets representing alternative solutions. These different spanning sets can be constructed and further evaluated (e.g. unification) in a totally independent way, thus eliminating the sometimes complicated organization overhead for the management of variable bindings occurring in many OR-parallel evaluation models for PROLOG. Furthermore, the separation of the spanning set construction and the unification phase leads to relatively large term structures, thus encouraging term or unification parallelism. Since there is relatively little experience with implementations based on spanning sets, certain doubts about the practicability and overall efficiency of this approach remain; but even if it is not good enough to be used as a self-contained evaluation mechanism, it may be used in combination with conventional ones to provide a better guidance through the search space.

---

[9] note that propositional validity is guaranteed for each spanning set according to its construction

# 4.9   Reductions

One of the main problems in the evaluation of logic formulae is the
traversal of the search space in order to find a derivation for it. Depending on the structure of the formula, this search space can be extremely huge, and actions to reduce it before searching can well pay off
in the end[10]. These reductions, however, have to maintain the essential logical characteristics of the formula under investigation. The reductions discussed here can be categorized into *equivalence*-preserving
and *satisfiability*-preserving transformations.  Equivalence-preserving
transformations typically operate on parts of the program only (single
clauses or sets of clauses), satisfiability-preserving reductions take into
account the whole program.  Examples for reductions are

- equivalence-preserving
  o tautology reduction: a clause contains a literal together with its
    negation;
  o subsumption:  a 'stronger' (more general) part of the program
    (clause) subsumes a 'weaker' (more specialized) one;
  o factorization: a clause subsumes a proper subset of itself;
  o restricted unit resolution: under certain conditions, a literal of a
    clause complementary to a unit clause can be deleted;
- satisfiability-preserving
  o pure literals: literals which do not have a unification partner in the
    formula;
  o isolated connections: pairs of complementary literals without con-
    nections to the rest of the formula.

The application of reductions, especially combinations of them, often
leads to considerably smaller search spaces to be traversed, and thus
to better evaluation times for a given formula; in some cases, reductions may even lead to a complete proof of certain formulae. On the
other hand, reductions only enfold their full potential if applied to a
formula with a certain complexity such that it is unfeasible, or at least
very difficult, to reduce the formula manually. For many typical theorem proving and logic programming problems this assumption does

---

[10] since these reductions – maybe together with other measures like a test for
propositional satisfiability – are performed prior to the actual processing of the
formula, they are often referred to as *preprocessing*, cf. (Letz et al., 1990a)

not hold, since often their complexity is not very high, or they are handcrafted because of their specific importance.

Reductions can be kept mutually independent, and work simultaneously on (instances of) one and the same formula. Furthermore, they often consist of basic operations performed on different parts of the formula. These two features makes them good candidates for the exploitation of parallelism; the restriction to this comes from the particular formula under evaluation, e.g. its complexity or its size.

The rest of this section contains brief descriptions of a number of different reduction methods; for further information consult for example (Letz et al., 1990a; Bibel, 1987; Robinson, 1965).

## Tautology Reduction

The basic idea of tautology reduction is to eliminate clauses with statements that are always true (tautologies); consider the case that a clause $C_i$, consisting of the set of literals $\{\mathcal{L}_{i1}, ..., \mathcal{L}_{in}\}$, contains a literal $L_j$ together with its negation $\neg L_j$: Obviously this clause cannot contribute to a proof of the formula, and hence it may be deleted. Although tautologies should not occur in carefully written programs, its underlying operation is a simple matching task, and basically can be done while the formula under evaluation is transformed into its internal representation.

**Program TAUTOLOGY**
**parameters**

| *name* | *type* | *description* |
|---|---|---|
| $F$ | set of clauses | formula |
| $C_i$ | set of literals | clause |
| $L_{ij}$ | literal | literal $j$ of clause $i$ |

**computation**

$\parallel\ \langle\ \ i : C_i \in F ::$      for all clauses
$\qquad C_i\ \ :=\ \ \{\}$ if $\exists j : (L_{ij}, L_{ik}) \in C_i ::$      remove tautological
$\qquad\qquad\qquad L_{ij} = \neg L_{jk}$      clauses
$\ \ \rangle$
**end TAUTOLOGY**

Figure 4.12: Program TAUTOLOGY

<table>
<tr><td colspan="3" align="center">TAUTOLOGY REDUCTION</td></tr>
<tr><td></td><td align="center">data structures</td><td align="center">operations</td></tr>
<tr>
<td>compu-<br>tation</td>
<td>input data<br>• clause $C_i$, consisting of the set of literals $\{\mathcal{L}_{i1}, ..., \mathcal{L}_{in}\}$<br>output data<br>• delete $C_i$ [yes/no]</td>
<td>matching<br>• check for the occurrence of a literal $\mathcal{L}_j$ together with its negation $\neg\mathcal{L}_j$</td>
</tr>
<tr>
<td>repre-<br>sen-<br>tation</td>
<td>data<br>• clause $C_i$, consisting of the set of literals $\{\mathcal{L}_{i1}, ..., \mathcal{L}_{in}\}$<br>evaluation support<br>• list of connections<br>• list of predicates</td>
<td>access to code<br>access to data<br>• delete clause $C_i$</td>
</tr>
<tr>
<td>com-<br>muni-<br>cation</td>
<td>not required (no shared data, no exchange of information)</td>
<td>not required</td>
</tr>
</table>

Table 4.10: Tautology Reduction

## Subsumption

A clause $C_1$ is said to *subsume* another clause $C_2$ iff there is a substitution $\sigma$ such that $\sigma(C_1) \in C_2$ (Robinson, 1965). Informally, this means that the second clause contains additional information (subgoals) to come to the same conclusion: it is more specialized than the first one. It is sufficient to keep the more general clauses and delete the subsumed ones.

Test for subsumption in a formula is a very costly action: first, each clause has to be compared with all others; second, the comparison operation requires to find a substitution satisfying certain conditions. Overall, this results in exponential explosion and presents a NP-complete problem. Subsumption can be optimized by performing it in two steps:

1. determine the set of *subsumption candidates* for each clause $C_i$; these are the clauses which contain (at least) all predicate symbols of $C_i$.

2. check subsumption of $C_i$ against each element of the previously computed set.

**Program SUBSUMPTION**
**parameters**

| name | type | description |
|---|---|---|
| $F$ | set of clauses | formula |
| $C_i$ | set of literals | literals of a clause |

**local variables**

| | | |
|---|---|---|
| $\sigma_i$ | variable bindings | substitution for clause $C_i$ |

**computation**

$$\|\ \langle\ \ i, j :\ C_i \in F, C_j \in F :: \qquad \text{for all pairs of clauses}$$
$$C_j\ :=\ \ \{\}\ \text{if}\ \exists\, \sigma_i :: \sigma_i(C_i) \in C_j \qquad \text{remove subsumed clauses}$$
$$\rangle$$

**end SUBSUMPTION**

Figure 4.13: Program SUBSUMPTION

| SUBSUMPTION | | |
|---|---|---|
| | data structures | operations |
| compu-tation | **input data**<br>• pair of clauses $(C_i, C_j)$;<br>clause $C_i = \{\mathcal{L}_{i1}, ..., \mathcal{L}_{in}\}$,<br>clause $C_j = \{\mathcal{L}_{j1}, ..., \mathcal{L}_{jn}\}$<br>**output data**<br>• delete $C_i$ [yes/no] | **subsumption:**<br>• check if $C_i$ subsumes $C_j$:<br>determine the set of clauses<br>which contain all predicates<br>occurring in $C_i$; does $C_i$<br>subsume these clauses? |
| repre-sen-tation | **data**<br>• clause $C_i = \{\mathcal{L}_{i1}, ..., \mathcal{L}_{in}\}$<br>• clause $C_j = \{\mathcal{L}_{j1}, ..., \mathcal{L}_{jn}\}$<br>**intermediate data**<br>• set of clauses which contain<br>all predicates of $C_i$<br>**evaluation support**<br>• predicate symbols $\mathcal{P}_i$ in $C_i$ | **access to code**<br>**access to data**<br>• delete clause $C_i$ |
| com-muni-cation | **implicit**<br>• shared data (formula)<br>**explicit**<br>• messages (deleted clauses) | **implicit**<br>• shared data access<br>**explicit**<br>• message passing |

Table 4.11: Subsumption

For this optimization, however, an appropriate internal representation of the formula under investigation is essential. Due to the rare occurrence and high cost of subsumption, usually a time limit is set to restrict the efforts bestowed on it.

## Factorization

A clause $C_i$ subsuming a proper subset $\hat{C}_i$ of itself can be substituted by $\hat{C}_i$; thus, subsumption subsumes factorization and the latter in fact is included in the above definition: it is not required that the two clauses to be checked for subsumption are different.

| FACTORIZATION | | |
|---|---|---|
| | data structures | operations |
| **compu-tation** | **input data**<br>• clause $C_i$, consisting of the set of literals $\{\mathcal{L}_{i1}, ..., \mathcal{L}_{in}\}$<br>**output data**<br>• replace $C_i$ by its subset $\hat{C}_i$ | **subsumption:**<br>• check if the clause $C_i$ subsumes a proper subset $\hat{C}_i$ of itself |
| **repre-sen-tation** | **data**<br>• clause $C_i$, consisting of the set of literals $\{\mathcal{L}_{i1}, ..., \mathcal{L}_{in}\}$<br>**intermediate data**<br>• set of clauses $\hat{C}_i$, consisting of the proper subsets of the clause $C_i = \{\mathcal{L}_{i1}, ..., \mathcal{L}_{in}\}$<br>**evaluation support**<br>• list of connections<br>• list of predicates | **access to code**<br>**access to data** |
| **com-muni-cation** | **implicit**<br>• shared data (clause)<br>**explicit**<br>• messages (clause, subsets) | **implicit**<br>• shared data access<br>**explicit**<br>• message passing |

Table 4.12: Factorization

Factorization especially may occur in synthetic calculi (e.g. resolution) in cases where clauses are (partially) instantiated during the computation of a proof. The incorporation of factorization actually can be viewed as a quite powerful extension of certain calculi, improving their strength with respect to the length of a proof considerably (Letz et al., 1990a); care must be taken, however, not to use factorization in a circular way in order to maintain soundness.

## Restricted Unit Resolution

If there exists a unifiable connection between a literal of a unit clause, and a literal of a normal clause, and the resolvent is a subset of that clause, the latter can be replaced by the resolvent. In this case a literal $L_j$ can be removed from a clause $C_n$ if there exists a literal $L_i$ with a complementary predicate symbol in a *unit* clause $C_m$, and a substitution $\sigma$ for the arguments of $L_i, L_j$ such that no substituted variable of $L_j$ appears in other literals of $C_n$.

**Program RESTRICTED UNIT RESOLUTION**
**parameters**

| *name* | *type* | *description* |
|---|---|---|
| $F$ | set of clauses | formula |
| $C_i$ | set of literals | clause |
| $L_{ik}$ | literal | literal $k$ of clause $C_i$ |
| $V_m$ | variable | variable in a literal |

**local variables**

| | | |
|---|---|---|
| $\sigma_{kl}$ | variable bindings | substitution |

**computation**

$$
\begin{aligned}
&\parallel \langle \quad i,j,k,l,m : C_i \in F, C_j \in F \qquad && \text{for all pairs of clauses} \\
&\qquad \wedge\ unit(C_i) && \text{with one unit clause,} \\
&\qquad \wedge\ L_{ik} \in C_i, L_{jl} \in C_j && \text{all pairs of literals of the clauses,} \\
&\qquad \wedge\ V_m \in L_{jl} :: && \text{all variables of the non-unit clause} \\
&\quad C_j := C_j \setminus Lk \ \text{if}\ \exists\, \sigma_{kl} :: && \text{remove subsumed clauses} \\
&\qquad\qquad \sigma_{kl}(L_{ik}) = \sigma_{kl}(L_{jl}) && \text{if there is a suitable substitution} \\
&\qquad\qquad \wedge\ \forall V_m \in \sigma_{kl}(L_{jl}) :: && \text{no substituted variable of } L_l \\
&\qquad\qquad V_m \notin C_j \setminus L_l && \text{is contained in other literals of } C_j
\end{aligned}
$$

$\rangle$

**end RESTRICTED UNIT RESOLUTION**

Figure 4.14: Program RESTRICTED UNIT RESOLUTION

| RESTRICTED UNIT RESOLUTION | | |
|---|---|---|
| | **data structures** | **operations** |
| **compu-tation** | **input data**<br>• pair of clauses $(C_m, C_n)$<br>**output data**<br>• clause $C_n \setminus L_j$ | **unit clause:**<br>• check if $C_m$ is a *unit clause*<br>**resolution:**<br>• check if there are literals $L_i \in C_m$ and $L_j \in C_n$ with complementary predicate symbols, and a unifier $\sigma$ for the arguments of $L_i$ and $L_j$ such that no substituted variable of $L_j$ is contained in other literals of $C_n$ |
| **repre-sen-tation** | **global data**<br>• pair of clauses $(C_m, C_n)$<br>• sets of literals $\{L_{m1}, ..., L_{mk}\}$, $\{L_{n1}, ..., L_{nl}\}$<br>• set of predicates $\{P_1, ..., P_p\}$<br>• sets of terms $\{T_{m1}, ..., T_{mk}\}$, $\{T_{n1}, ..., T_{nl}\}$<br>**intermediate data**<br>• set of unifiers $\{\sigma_1, ..., \sigma_s\}$ | **access to code**<br>**access to data** |
| **com-muni-cation** | **implicit**<br>• shared data (clauses)<br>**explicit**<br>• messages (clauses) | **implicit**<br>• shared data access<br>**explicit**<br>• message passing |

Table 4.13: Restricted Unit Resolution

## Purity Reduction

A literal $L_i$ of a formula $F$ is called *pure* if there is no connection to it, i.e. if there is no (complementary and) unifiable literal $L_j$ in a different clause of $F$. Clauses containing pure literals cannot contribute to the proof of a formula, and thus may be deleted. This removal of clauses may again give rise to new pure literals, entailing repetitive applications of purity reduction.

**Program** PURITY REDUCTION
**parameters**

| *name* | *type* | *description* |
|---|---|---|
| $F$ | set of clauses | formula |
| $C_i$ | set of literals | clauses of the formula |
| $L_{ij}$ | literal | literal of clause $C_j$ |
| $CONN$ | set of connections | connections in the formula |

**computation**

$$\parallel \; \langle \;\; i,j : C_i \in F, L_{ij} \in C_i :: \qquad \text{for all clauses and literals}$$
$$C_i \;\; := \;\; \{\} \text{ if } \exists\, L_{ij} \in C_i :: \qquad \text{remove clauses}$$
$$L_{ij} \notin CONN \qquad \text{with pure literals}$$
$$\rangle$$

**end** PURITY REDUCTION

Figure 4.15: Program PURITY REDUCTION

| PURITY | | |
|---|---|---|
| | **data structures** | **operations** |
| **compu-tation** | **input data**<br>• clause $C_i = \{\mathcal{L}_{i1}, ..., \mathcal{L}_{in}\}$<br>• set of connections $CONN$<br>**output data**<br>• delete $C_i$ [yes/no] | **matching**<br>• check for the occurrence of a literal in the set of connections:<br>$\exists\, L_{ij} \in C_i : L_{ij} \notin CONN$ |
| **repre-sen-tation** | **global data**<br>• set of connections $CONN$<br>• clause $C_i = \{\mathcal{L}_{i1}, ..., \mathcal{L}_{in}\}$<br>**evaluation support**<br>• set of predicates | **access to code**<br>**access to data**<br>• remove clause $C_i$ if it contains a pure literal |
| **com-muni-cation** | **implicit**<br>• shared data (clauses)<br>**explicit**<br>• messages (clauses) | **implicit**<br>• shared data access<br>**explicit**<br>• message passing |

Table 4.14: Purity

The program given in Figure 4.15 actually describes a somewhat restricted purity reduction (which might be called *weak purity*): it only checks if a literal is not contained in the set of connections[11] for that formula; "strong" purity would have to check (full) unifiability as well.

## Isolated Connection Reduction

Extending the concept of purity leads to isolated literals: A *literal $L_{ik}$* in a clause $C_i$ of $F$ is called *isolated* iff there is exactly one (unifiable) connection $CONN_{ik,jl} = (L_{ik}, L_{jl})$ literal $L_{ik}$ is part of[12]; a *connection* is called *isolated* if an isolated literal is part of a unit clause, or if both literals of a connection are isolated.

**Program** ISOLATED CONNECTION REDUCTION
**parameters**

| name | type | description |
|------|------|-------------|
| $F$ | set of clauses | formula |
| $C_i$ | set of literals | clauses of the formula |
| $L_{ik}$ | literal | literal $k$ of clause $C_j$ |
| $CONN$ | set of connections | connections in the formula |
| $CONN_{ik,jl}$ | connection | connection between $L_{ik}$ and $L_{jl}$ |

**computation**

| | |
|---|---|
| $\Vert\ \langle\ \ i,j,k,l:\ C_i,C_j, \in F, L_{ik} \in C_i, L_{jl} \in C_j ::$ | for all pairs of clauses and their literals |
| $\qquad C_i := resolvent(L_{ik}, L_{jl})$ | replace clauses by the resolvent |
| $\qquad\qquad$ if $L_{ik} \in C_i\ \wedge\ L_{ik} \in CONN$ | if the connection |
| $\qquad\qquad \wedge\ \lvert CONN_{ik,jl}\rvert = 1$ | is isolated |
| $\rangle$ | |

**end** ISOLATED CONNECTION REDUCTION

Figure 4.16: Program ISOLATED CONNECTION REDUCTION

The measure to be taken, however, is not simply a removal of the isolated literal or the clause it is part of: a resolution step is performed, and the clause containing the isolated literal is replaced by the resolvent. As a result, other literals may become isolated as well, potentially resulting again in an iterative application; in some cases,

---

[11] it is assumed that only connections between weakly unifiable literals appear
[12] if clauses $C_i$ and $C_j$ are the same, then an isolated literal is also pure

| ISOLATED CONNECTION REDUCTION | | |
|---|---|---|
| | **data structures** | **operations** |
| **compu-tation** | **input data**<br>• clause $C_i$, consisting of the set of literals $\{\mathcal{L}_{i1}, ..., \mathcal{L}_{in}\}$<br>• set of connections $\mathcal{CONN}$<br>**output data**<br>• replace $C_i$ by the resolvent $(L_{ik}, L_{jl})$ | **matching**<br>• check for the occurrence of a literal in the set of connections:<br>$L_{ik} \in C_i \wedge L_{ik} \in \mathcal{CONN}$<br>• exactly one connection satisfying the above condition<br>**unification**<br>• compute the resolvent $(L_{ik}, L_{jl})$ |
| **repre-sen-tation** | **global data**<br>• set of connections $\mathcal{CONN}$<br>• clause $C_i$, consisting of the set of literals $\{\mathcal{L}_{i1}, ..., \mathcal{L}_{in}\}$<br>• resolvent $(L_{ik}, L_{jl})$<br>**evaluation support**<br>• list of predicates | **access to code**<br>**access to data**<br>• replace $C_i$ by the resolvent $(L_{ik}, L_{jl})$ |
| **com-muni-cation** | **implicit**<br>• shared data (clause, set of connections)<br>**explicit**<br>• messages (clause) | **implicit**<br>• shared data access<br>**explicit**<br>• message passing |

Table 4.15: Isolated Connection Reduction

even a proof of the formula may be found[13].

The reductions described above transform the syntactical appearance of a formula while leaving its main semantical properties unchanged. As a result of such a reduction the formula might be in a form where again a reduction (the same or a different one) can be applied. In some cases, this can lead to a snowball effect reducing the search space to be traversed drastically.

---

[13] this is due to the fact that with resolution a step is performed which can be the basis of a proof mechanism

### Termination and Correctness

This program is guaranteed to reach a fixed point where all possible reductions have been applied, and the formula $F'$ resulting from these transformations is satisfiable iff the original formula $F$ is satisfiable. Each application of a transformation results in a decreased size of the formula, either by removing some parts (literals, clauses), or by a resolution step which combines two clauses into one clause; each transformation maintains equivalence or satisfiability, and thus any possible sequence of transformations preserves these characteristics as well.

Table 4.16 gives an overview of the features common to all reductions investigated here; these reductions, especially with their potential to be applied concurrently to a formula as well as parallelized internally, are investigated in more detail in the following.

### Combinations

In most cases, reductions will be applied in a preprocessing phase before the actual proof of a formula starts since they can be used to reduce the search space to be traversed considerably.

**Reductions - Spanning Sets**   Reductions can be integrated favorably into the very first phase of the spanning set approach when the connections appearing in the formula are determined. Many of the operations and intermediate results required for this purpose are used by (some) reductions as well, and the profit gained through the application of reductions is already visible during the generation of spanning sets. An application of reductions in later stages is not very useful; most modifications achieved through reductions are made obsolete by the directed construction of spanning sets of connections.

**Reductions - Competition**   If reductions are applied independently to one and the same formula, and then a proof of the reduced formula is computed, this application can be viewed as a certain kind of competition. The same holds for combinations or particular sequences of reductions which go together well. Some kinds or combinations of reductions also are particularly suited together with some specific evaluation mechanisms; the ones described in this section, for example, are used in the preprocessing part of SETHEO and PARTHEO (Ertel et al., 1989).

| REDUCTIONS | | |
|---|---|---|
| | **data structures** | **operations** |
| **compu-<br>tation** | **input data**<br>• logical formula (program)<br>**output data**<br>• reduced logical formula (program)<br>**intermediate data**<br>• sets of clauses<br>• sets of literals<br>• sets of predicates<br>• substitutions | **reduction mechanisms**<br>• equivalence-preserving<br>• satisfiability-preserving<br>• global (works on the whole formula)<br>• local (works on parts of the formula)<br>**basic operations**<br>• matching<br>• unification<br>• subsumption<br>• resolution |
| **repre-<br>sen-<br>tation** | **reduction mechanisms**<br>• program code of the single reduction mechanisms<br>**data**<br>• logical formula (program)<br>**intermediate data**<br>• sets of clauses<br>• sets of literals<br>• sets of predicates<br>• substitutions<br>**administrative data**<br>• addresses of cooperating reduction mechanisms<br>• activities of other reduction mechanisms | **access to code**<br>**access to data** |
| **com-<br>muni-<br>cation** | **implicit**<br>• parameters<br>• shared data structures<br>**explicit**<br>• messages | **implicit**<br>• (remote) procedure call<br>• shared data access<br>**explicit**<br>• message passing |

Table 4.16: Reductions

**Reductions - Precision**   Reductions, especially the more complicated types, can also be applied with various precision, again ranging from propositional logic to predicate logic with full unification. Appropriate types of reductions can also be combined with evaluation mechanisms of different precision.

**Reductions - Modularity**   Especially in the context of reusable software and meta-programming techniques, reductions can be made available as standard modules and integrated into specific user-defined evaluation mechanisms. On the other hand, reductions may be applied to the single modules of large, modular programs in parallel.

**Reductions - Multitasking**   Although the application of reductions consists of largely independent computations, in most cases the grain size of the computation tasks is too small for multitasking.

## Conclusions

Reductions present a very effective tool for a reduction of the search space to be traversed by an evaluation mechanism for logical formulae. In addition, the different kinds of reductions can be combined in a number of ways without greater problems, often resulting in mutual improvements. Their potential for parallelism is quite large, due to the independence of the basic operations which often can be applied to different parts of a formula independently. Experiments with parallel implementations of the reductions described above, however, have shown that the overhead for parallelization usually is way too high to justify the potential gain of performance with respect to a sequential implementation. On the other hand, the formulae used for these experimentations are typical theorem-proving problems: carefully coded and rather small. For formulae of a size which is too large to be optimized manually, or for (semi-)automatically generated programs, reductions have the potential to reduce the computational complexity of the evaluation of a formula from practically unfeasible to easily manageable on a normal workstation. Reductions also present a way to build optimizing compilers for logic programming languages, and there are far more equivalence- or (un-)satisfiability preserving transformations which can be used for that purpose than could be mentioned here.

# 4.10   OR-Parallelism

Similar to the spanning set approach described in Section 4.8, OR-parallelism aims at a simultaneous exploration of alternatives in the search space. In contrast to the mainly static analysis of the formula used with spanning sets, in an OR-parallel execution scheme different alternatives are initiated dynamically as they occur. This is the case in situations where more than one clause can be used for extension of the (sub-)goal under evaluation, i.e. the literal representing the current goal is complementary and unifiable to more than one literal representing the head of a clause. Investigating different alternatives in parallel is a way to avoid backtracking which is used in the sequential case when an alternative fails.

An abstract execution model for OR-parallelism can be based on the construction of trees representing proofs of the formula to be evaluated; for each alternative solution, a separate tree exists, but trees representing identical proofs may be created independently. In the following an algorithm is presented which describes the exploitation of OR-parallelism based on the model elimination calculus (cf. Section 2.3). This program specifies only an abstract execution scheme for OR-parallelism and is not intended for direct implementation due to the abundance of information which it implies.

## Problem Description

A proof for a formula is to be found, checking potential alternative solutions in parallel. In clause notation, alternative solutions arise if there are more than one clauses with the same head literals.

## Design of a Solution

The central idea is to perform for a (sub-) goal as many alternative extension steps as there are clause heads complementary to and unifiable with this subgoal. The body of a clause, consisting of a set of subgoals, together with its environment (variable bindings, trail, depth) is regarded as an OR-task, and during an extension step new OR-tasks are created whenever there are multiple extensions for a subgoal.

The current environment of an OR-task is not detailed in the above program since its definition depends on the scope; minimally it contains just enough information to (deterministically) redo the proof computed so far (see *reproof* below), maximally the whole informa-

**Program OR**
**parameters**

| name | type | initially | description |
|---|---|---|---|
| $F$ | set of clauses | | formula to be proven |

**local variables**

| | | | |
|---|---|---|---|
| $\mathcal{ORTS}$ | set of OR-tasks | {goal clause} | tasks to be evaluated |
| $ORT_i$ | $(clause, envir)$ | {} | single OR-Task |
| $\mathcal{CAND}_{ij}$ | set of clauses | {} | candidate clauses for extension of $ORT_i$ with subgoal $j$ |
| $C_k$ | set of literals | {} | clause of the formula |
| $head(C_k)$ | literal | {} | head literal of clause $C_k$ |
| $subgoal_{ij})$ | literal | {} | subgoal $j$ of $ORT_i$ |
| $\sigma_{ijk}$ | variable bindings | {} | substitutions for the unification of the head of clause $C_k$ with $subgoal_j$ of $ORT_i$ |
| $envir_{ijk}$ | variable bindings, trail, ... | {} | important information about the proof process |

**invariants**

| | |
|---|---|
| $\sigma_{ik}(head(C_k)) = \sigma_{ik}(goal(ORT_i))$ | head of clause $C_k$ and goal of $ORT_i$ are unifiable with substitution $\sigma_{ik}$ |

**computation**

```
[]  ⟨ i, j : ORTᵢ ∈ ORTS, subgoalᵢⱼ ∈ ORTᵢ ::       for all OR-Tasks and
                                                      their subgoals

      []  ⟨ k : Cₖ ∈ F ::                            for all clauses
          CANDᵢⱼ := {(Cₖ, σᵢⱼₖ) :                    collect
                    σᵢⱼₖ(head(Cₖ)) = σᵢⱼₖ(subgoalᵢⱼ)}  candidate clauses
          ⟩
      []  ⟨ l : (Cᵢⱼₗ, σᵢⱼₗ) ∈ CANDᵢⱼ ∧ l > 1 ::     for all alternative candi-
                                                      date clauses
          ORTS  := ORTS ∪ (Cᵢⱼₗ, envirᵢⱼₗ)           create new OR-tasks
          ⟩
    ⟩
end OR
```

Figure 4.17: Program OR

tion computed so far is stored and copied each time a new OR-task is created (see *copying* below).

## Correctness and Complexity

The program OR (Figure 4.17) reaches a fixed point if no more new OR-tasks can be created; this is the case when no more alternatives are encountered, or encountered alternatives are not compatible with the current environment (i.e. unification fails). Due to the possibility of cycles (recursion) it cannot be guaranteed that a fixed point is reached. Note that this program deals only with the management of OR-parallelism; thus, no statement is made about the way the proofs for the single alternatives are obtained.

## Refinements

The main problem in the OR-parallel execution of logic programs lies in the management of previously computed data (history), especially the variable bindings produced so far, together with the goals used and their sequence. The most important management task is to provide access to the history and environment of a proof approach; it also can take care of optimizations based on "pruning" of dead branches. In addition, since the number of available processing elements typically is much smaller than the number of tasks produced by OR-parallelism, a task scheduling mechanism has to be provided. Implementation mechanisms for the history management of OR-parallel proof approaches are:

- Copying: All important data are copied for each alternative proof approach (tree); this is of course very expensive, both in terms of communication required and memory used.

- Reproof: Only an outline of the computation required to reach a particular point in the execution is transferred and used to re-do the proof; this outline can be very compact (a few bytes), and the extra computation needed is very little since it is fully deterministic (Schumann and Letz, 1989).

- Binding lists: The bindings produced so far are represented via lists accessible to all relevant proof approaches; the access to very 'old' bindings, of course, is very slow here because the whole list has to be traversed for dereferencing.

- Cactus stacks: As an extension to the stacks used in the Warren Abstract Machine, an entry of a stack may consist of a pointer to another stack, resulting in so-called 'cactus stacks'.

- Binding arrays: Constant-time access to binding information is provided by binding arrays, indexed via the depth of the tree; this requires information on which level of the tree a certain variable has been bound.

- Hash windows: The selection of binding information for a certain variable is done through a hash function.

- Versions vector: The elements of a central versions vector represent the current variable bindings. Drawbacks here are: synchronization problems on variable access, shared data belong to various processors are not protected against modifications by unauthorized processors, and the model is based on a fixed number of processors corresponding to the vector length.

Most of the implementations of PROLOG using OR-parallelism use combinations or modifications of cactus stacks and binding arrays (Haridi and Brand, 1988; Lusk et al., 1988; Warren, 1987; Westphal et al., 1987). All these approaches, however, are based on (partially) shared memory systems; PARTHEO is based on a distributed implementation with OR-parallelism and reproofs.

### Internal Parallelization

The behavior of a system based on OR-parallelism is of a very dynamical, execution-oriented nature: new computational units (tasks) are created and executed at runtime, possibly again creating new ones. This leads to a tree-like task structure with a maximum amount of parallelism determined by the respective number of extension candidates for a subgoal. In practice, not the full potential is exploited since in some cases the computation to be performed in a new task is too little to justify the overhead involved[14], and it is much more efficient to do it sequentially. In addition, further internal parallelization can be achieved in dependence of the actual evaluation mechanism.

### Combinations

Due to its highly dynamical and execution-oriented nature, OR-parallelism cannot be combined easily with some of the categories described in the previous sections. In many cases, the overhead for the management of parallelism introduced through these combinations is too high for practical usability. On the other hand, some of the problems occurring with OR-parallelism can be reduced through combinations with other categories (e.g. Spanning Sets, Precision, Multitasking).

---

[14] typical examples are the termination conditions of recursive 'procedures'

| OR-PARALLELISM | | |
|---|---|---|
| | **data structures** | **operations** |
| **computation** | **input data**<br>• formula<br>**output data**<br>• overall proof result<br>• alternative proofs<br>• proof information (proof tree, substitutions)<br>**intermediate data**<br>• alternative tasks<br>• environment (clause, subgoal extension clauses, variable bindings)<br>**evaluation support**<br>• reproof information, or<br>• binding lists / arrays, cactus stacks, hash windows | **task evaluation**<br>• evaluation mechanism for single tasks<br>**task creation**<br>• creation mechanism for new tasks<br>**task management**<br>• task distribution<br>• load balancing<br>• stop evaluation |
| **representation** | **program:** code for<br>• task evaluation<br>• task creation<br>• task management<br>**global data**<br>• formula<br>• proof(s)<br>**local data**<br>• alternative tasks<br>• environment<br>**evaluation support**<br>• reproof information<br>• variable bindings | **access to code**<br>**access to data structures**<br>**evaluation support**<br>• hash functions<br>• (cactus) stack management<br>• (distributed) garbage collection<br>• memory management (for multiple tasks on a node) |
| **communication** | **implicit**<br>• parameters<br>• shared data structures (cactus stacks, hash windows)<br>**explicit**<br>• messages (reproof info) | **implicit**<br>• parameter passing mechanisms, e.g. (remote) procedure call<br>• shared data access<br>**explicit**<br>• message passing |

Table 4.17: OR-Parallelism

**OR-parallelism – Reductions**   As in the general case, the application of reductions with the potential decrease of the search space is well suited for a combination with OR-parallelism. In addition to its application before the actual evaluation, some reduction mechanisms (e.g. factorization) can be incorporated into the evaluation mechanism to check the newly created sub-formula (task) for possible reductions.

**OR-parallelism – Spanning Sets**   OR-parallelism and Spanning Sets are two complementary approaches to the computation of alternative solutions: the first is dynamical and needs good execution support, whereas the latter is based on a statical analysis of the program, leading to independent computational units. Nevertheless they can be combined with mutual benefits: First, a simple statical analysis (which does not have be as thorough as the one of (Ibañez, 1989)) is performed to identify the initial alternative proof attempts, which are completely independent; then these proof attempts are evaluated in a (dynamical) OR-parallel way. This combination avoids the overhead for the management of cycles in the spanning set approach, and reduces the complications involved with the administration of variable bindings in OR-parallelism.

**OR-parallelism – Competition**   Since the basic principle of OR-parallelism is to compute alternative proofs for one and the same formula, it represents an example for a competitive approach. This direction can be further sophisticated by applying different evaluation mechanisms to newly created tasks, although in practice the overhead involved will probably be too high.

**OR-parallelism – Precision**   In principle, all newly created tasks can be tackled by evaluation mechanisms incorporating different precision; the overhead involved here is considerable, especially since the creation of new tasks has to be coordinated between the instances of tasks evaluated with different precision. A useful limitation is to investigate the possible creation of new tasks with low precision (e.g. propositional logic), aiming at a well-founded decision about their parallel or sequential execution.

**OR-parallelism – Modularity**   A useful combination is to evaluate single modules in an OR-parallel way. While OR-parallelism might occur across boundaries of modules (depending on the way they are defined), its utilization can lead to substantial overhead.

**OR-parallelism – Multitasking**  As already indicated by the terms used throughout this section, OR-parallelism and multitasking represent an excellent combination: OR-parallelism provides the framework for the evaluation of the formula, and multitasking the environment for the execution at run-time.

### Conclusions

OR-parallelism is the most popular category of parallelism exploited in logic programming. It can be implemented with relatively little effort by modifying existing evaluation mechanisms, and leads to considerable performance gain in many applications (Chassin de Kergommeaux et al., 1988). In contrast to the spanning set approach, where a statical analysis leads to completely independent proof approaches, the dynamic task creation of OR-parallelism requires precautions to maintain the compatibility of variable bindings between father- and son-tasks.

In parallel systems with a global address space (usually via shared memory), a number of techniques have been developed for that purpose. For systems without a global address space (e.g. Transputer-based machines), these techniques are not applicable; in PARTHEO, a *reproof* concept has been implemented where the 'old' variable bindings are computed again via a re-evaluation of the previous task. In contrast to the first impression, this does not result in a large amount of redundant work: first, the computation to be redone is deterministic, and second, the information to be transferred for the creation of a new task is extremely small (a few bytes, typically). Very good results have been achieved with this concept in the PARTHEO system (Schumann and Letz, 1990; Ertel et al., 1989).

# 4.11   Routes

This method of exploiting parallelism is based on a (statical) analysis of the formula to be evaluated; in this analysis, certain 'routes' of connections[15] may be identified which have to be traversed independent of the actual parameter values used. In terms of the connection method, a route represents a partial path through the matrix covered by connections, which contains OR-literals[16] only at its starting or end points; in terms of computation, a route represents a sequence of operations which can be performed deterministically.

The crucial point for the introduction of parallelism in the evaluation of routes of connections is to get away from the *sequence* of operations: there is actually no need to resolve such a route sequentially with an execution time linear to the number of components. The components of a route can as well be treated pairwise in parallel, resulting in logarithmic execution time (provided that there is a sufficient number of processing elements).

In a sense routes have a similar relation to AND-parallelism as the spanning set approach to OR-parallelism: routes and spanning sets are based on a *statical* analysis of a program leading to computational segments without or only with well-defined interactions, whereas AND- and OR-parallelism are invoked dynamically with a complicated management of interactions.

The static derivation of routes and spanning sets has also a serious drawback, which lies in the occurrence of *cycles* representing recursive definitions of predicates. Many typical instances of cycles, however, can be handled reasonably well, restricting the management required dynamically during execution considerably; for an elaboration see (Ibañez, 1989; Balcke, 1987). In addition, theoretical results from recursion theory (Schnorr, 1974), complexity theory (Paul, 1978; Parberry, 1987) as well as transformation methods produced for conventional languages (Bauer and Wössner, 1981; Brandes, 1988) may be transferred; the treatment of simple cases (e.g. tail recursion) can already be found in implementations of logic programming languages (Sterling and Shapiro, 1986).

---

[15] which during the evaluation are tracked by *extension* or *resolution* steps, depending on the calculus used

[16] literals for which alternative connections exist

### Problem Description

Based on a statical analysis a formula is to be evaluated by an identification of *routes*, and the unification of the elements occurring in the particular routes.

### Design of a Solution

In the following specification of routes two aspects are covered: one is the treatment of different routes occurring in a program, the other unification of the elements (connected literals) in a route. The different routes are constructed by incrementally adding those literals to a route which are in a connection with a literal from the route.

### Program ROUTES

**parameters**

| name | type | initially | description |
|------|------|-----------|-------------|
| $F$ | set of clauses | | formula ( matrix) |
| $SPASE$ | set of connections | | spanning set |
| $CONN_i$ | pair of literals | {} | connection of the set |

**local variables**

| | | | |
|------|------|-----------|-------------|
| $ROUTES$ | set of routes | {} | all routes |
| $ROUTE_i$ | set of literals | $CONN_i$ | one route |
| $\sigma_i$ | variable bindings | nil | substitution for route $i$ |

**invariants**

$\forall i,j:\ ROUTE_i, ROUTE_j \in ROUTES ::$     all routes
$\qquad ROUTE_i \neq ROUTE_j$     are different

$\forall i,k:\ L_{ik} \in ROUTE_i, L_{ik} \in CLAUSE_k$     each literal of a route
$\qquad \exists l:\ L_{il} \in ROUTE_i \wedge L_{il} \in CLAUSE_l$     has a partner literal
$\qquad\qquad \wedge\ L_{il} \in CLAUSE_l$     in a different clause, and
$\qquad\qquad \wedge\ (L_{ik}, L_{il}) \in SPASE$     the two literals form a

    connection

$\forall k,l:\ L_{ik}, L_{il} \in ROUTE_i ::$     for all literals of a route
$\qquad \sigma(L_{ik}) = \sigma(L_{il})$     variable bindings must

    be consistent

**computation**

$\|\ \langle\ \ i,k,l:\ ROUTE_i \in ROUTES,$     for all routes
$\qquad\qquad L_{ik} \in ROUTE_i,$     and their literals
$\qquad\qquad (L_{ik}, L_l) \in SPASE ::$
$\qquad ROUTE_i := ROUTE_i \cup L_l$     add connected literal
$\quad \rangle$

**end ROUTES**

Figure 4.18: Program ROUTES

Unification is performed implicitly via an invariant, guaranteeing that only literals with compatible variable bindings are added to a route. The consistency of bindings for shared variables occurring in different roads is also maintained through an invariant. In order to avoid unnecessary complications, only a single spanning set of connections for a formula is regarded; copies of clauses whose necessity can be determined statically are assumed to be made, but cycles are not resolved. As a consequence, no alternatives (OR-parallelism) occur. The objective then is for each route to identify a most general unifier for the terms of the literals in the route, additionally satisfying the restrictions imposed by the occurrence of variables shared between subgoals (and thus different routes).

### Refinements

In analogy to the spanning set approach, a separation between the construction of routes and the unification of their literals can be made. This leads to more complex terms in the unification since sets instead of pairs of terms have to be unified.

### Combinations

The concept of routes is based on a statical analysis of the formula under investigation. Thus it can be combined more easily with related approaches like Spanning Sets, whereas it is more difficult to bring it together with dynamical ones like OR-parallelism. On the other hand, statical analysis alone is not sufficient for large classes of formulae, and has to be extended towards dynamical execution, e.g. to cope with recursion.

**Routes – OR-parallelism**   At a first glance the combination of routes and OR-parallelism doesn't make much sense due to their different execution behavior. A second look reveals potential mutual benefits, however: the routes concept helps to identify parts of the OR-parallel proof process where sets of literals can be unified instead of pairs. On the other hand, methods related to OR-parallelism must be applied in the cases where statical analysis does not suffice.

**Routes – Reductions**   In the case that routes are investigated after the computation of minimal spanning sets, reductions are not necessary. Otherwise the application of reductions before the actual evaluation is useful to decrease the amount of work during evaluation.

<table>
<tr><td colspan="3" align="center">ROUTES</td></tr>
<tr><td></td><td align="center">data structures</td><td align="center">operations</td></tr>
<tr><td>compu-<br>tation</td><td>input data<br>• formula<br>• (spanning) set of connections<br>output data<br>• proof<br>intermediate data<br>• routes<br>• substitutions</td><td>construction of routes<br>• set union<br>unification<br>• computation of variable bindings</td></tr>
<tr><td>repre-<br>sen-<br>tation</td><td>program<br>• code for route construction, unification<br>global data<br>• formula<br>• set of connections<br>local data<br>• routes<br>• substitutions</td><td>access to code<br>access to data</td></tr>
<tr><td>com-<br>muni-<br>cation</td><td>implicit<br>• shared data structures<br>explicit<br>• messages</td><td>implicit<br>• shared data access<br>explicit<br>• message passing</td></tr>
</table>

Table 4.18: Routes

**Routes – Spanning Sets** These two concepts provide a good mutual supplement: Spanning sets are a good basis for the identification of routes, and routes are useful to structure the information gained from the computation of spanning sets into largely independent[17] units of computation.

**Routes – Competition** The routes concept already presents a sophisticated version of a particular calculus[18], and does not leave much leeway for the use of different calculi, or the application of dif-

---

[17] except for the occurrence of shared variables
[18] or at least a class of analytical calculi

ferent strategies. The only possibilities are to proceed in various ways during the construction of routes, and to use various unification mechanisms.

**Routes – Precision**   One of the basic operations in the routes approach is unification; this leads directly to the use of unification mechanisms with different precision and provides a method to check out the viability of routes by using lower precision scout processes.

**Routes – Modularity**   The computational units appearing in the routes approach can be mapped onto modules: they are largely independent, but require some exchange of information for the consistency of shared variable bindings. The other way round, different modules of a logical formula can be evaluated in parallel according to the routes approach.

**Routes – Multitasking**   Apart from the more or less trivial combination to evaluate multiple tasks in parallel using the routes approach (together with spanning sets, for example), there does not seem to be much potential for mutual benefits. Routes require the exchange of information and also are too small to make good use of multitasking facilities, and routes alone are not sufficient to be used as evaluation mechanism for tasks consisting of logical programs.

**Conclusions**

Especially in combination with the spanning set approach, routes represent a useful method to structure a logical formula into largely independent units of computation, determined by a statical analysis of the formula. A restriction is imposed by the fact that for some classes of formulæ a statical analysis is not sufficient, and dynamic features (e.g. the creation of clause copies) must be integrated. This restriction is not so much of a fundamental nature but leads to a complicated management of information like variable bindings.

# 4.12  AND-Parallelism

The aim of AND-parallelism is to concurrently evaluate the subgoals
that make up the body of a clause. In contrast to OR-parallelism,
where the alternative clauses treated in parallel represent distinct
proof approaches, and where it is sufficient if one of them results in
a proof, in AND-parallelism all the subgoals of a clause – once the
clause is used – must be evaluated. Thus all the results of subgoals
evaluated in AND-parallel must be collected, and the evaluation of a
clause is successful only if all the subgoals succeeded. A problem may
arise here from the potential occurrence of variables shared between
subgoals, whose bindings must be consistent.

**Problem Description**

Try to find a proof for a formula $F$ by concurrently evaluating the sub-
goals $L_{i1}, \ldots, L_{in}$ of the current clause $C_i = \{L_{i0}, L_{i1}, \ldots, L_{in}\}$ while
maintaining the consistency of variables shared between subgoals.

**Design of a Solution**

The program AND/OR (Figure 4.19) is an extension of the one describing OR-parallelism. Additional AND-tasks are introduced, consisting
of the subgoals of a clause body to be evaluated; thus an OR-task
comprises a set of AND-tasks, each of which can in turn evoke a set of
OR-tasks, and so forth. The consistency of variable bindings is maintained through an invariant claiming that the variable substitutions
for all subgoals of one clause (all AND-tasks within one OR-task) be
compatible. This invariant, of course, is only a statement expressing
the necessity, without saying how to really do it; the management of
shared variable bindings is one of the main problems for implementations combining AND- and OR-parallelism.

**Correctness and Complexity**

The behavior of the program AND/OR (Figure 4.19) is quite similar to
the program OR (Figure 4.17): a fixed point is reached as soon as no
more (AND- or OR-) tasks can be created, but due to possible cycles
it cannot be guaranteed that a fixed point will be reached eventually.

The main problems concerning the implementation of AND-parallelism lie in the treatment of shared variables and the management of
nondeterminism (i.e. the combination of AND- and OR-parallelism).
A number of approaches have been pursued to deal with this problem:

**Program AND/OR**
**parameters**

| name | type | initially | description |
|---|---|---|---|
| $F$ | set of clauses | | formula to be proven |

**local variables**

| | | | |
|---|---|---|---|
| $\mathcal{ORTS}$ | set of OR-tasks | {goal clause} | tasks to be evaluated |
| $ORT_i$ | $(\mathcal{ANDTS}, envir_i)$ | {} | single OR-Task |
| $\mathcal{ANDTS}$ | set of AND-tasks | {subgoals of the goal clause} | tasks to be evaluated in an AND-parallel way |
| $ANDT_{ij}$ | $(subgoal_{ij}, envir_{ij})$ | {} | AND-Task |
| $\mathcal{CAND}_{ij}$ | set of clauses | {} | candidate clauses |
| $C_k$ | set of literals | {} | clause of the formula |
| $head(C_k)$ | literal | {} | head literal of clause $C_k$ |
| $subgoal_{ij})$ | literal | {} | subgoal $j$ of $ORT_i$ |
| $\sigma_{ijk}$ | variable bindings | {} | substitutions |
| $envir_{ijk}$ | variable bindings, trail, ... | {} | important information about the proof process |

**invariants**

$$\sigma_{ik}(head(C_k)) = \sigma_{ik}(goal(ORT_i))$$

head of clause $C_k$ and goal of $ORT_i$ are unifiable; substitution $\sigma_{ik}$

$$\forall i, j, k : ORT_i \in \mathcal{ORTS},$$
$$ANDT_{ij} \in ORT_i,$$
$$ANDT_{ik} \in ORT_i ::$$
$$\sigma_{ij}(subgoal_{ij}) = \sigma_{ik}(subgoal_{ik})$$

for all OR-Tasks, all pairs of AND-Tasks maintain consistency of variable bindings

**computation**

⫴ ⟨ $i, j : ORT_i \in \mathcal{ORTS},$           for all OR-Tasks and their AND-Tasks

  ⫴ ⟨ $k : C_k \in F ::$           for all clauses
     $\mathcal{CAND}_{ij} := \{(C_k, \sigma_{ijk}) :$           collect candidate clauses
          $\sigma_{ijk}(head(C_k)) = \sigma_{ijk}(subgoal_{ij})\}$
  ⟩

  ⫴ ⟨ $l, m : (C_{ijl}, \sigma_{ijl}) \in \mathcal{CAND}_{ij},$           for all candidate clauses
          $subgoal_{ijlm} \in C_{ijl} ::$
    ‖ $ORTS$     $:= ORTS \cup$           create new OR-tasks for
                  $(C_{ijl}, envir_{ijl})$ if $l > 1$           *alternative* cand. clauses
    ‖ $\mathcal{ANDTS}$     $:= \mathcal{ANDTS} \cup$           create new AND-tasks
                  $(subgoal_{ijlm}, envir_{ijlm})$
  ⟩
⟩

**end AND/OR**

Figure 4.19: Program AND/OR

- Explicit AND-Parallelism: The goals which may be executed in parallel are explicitly marked by the programmer, e.g. by a *parallel connective* ('//' or '|') in contrast to the sequential one (','). This approach leaves it completely to the user to decide where parallelism is to be introduced, as well as the consistency maintenance of shared variables (Moniz Pereira et al., 1988).

- Independent AND-Parallelism: Only goals without shared variables (or with restricted use of shared variables) may be treated in parallel; the task of identifying 'dangerous spots' can be done by the compiler so that there is no need for annotations (de Groot, 1984).

- Annotated Variables: The management of shared variables can be supported by indicating their intended use (e.g. as read-only).

- Consistency Check: The bindings of variables shared between subgoals are checked after the evaluation of the subgoals for consistency; if necessary (and possible), unification is performed to find a consistent binding.

- Determinate AND-Parallelism: In addition to independent goals, dependent goals may be executed in parallel as soon as they become determinate; a goal is determinate if there is at most one candidate clause for its extension[19] (Haridi and Brand, 1988).

- Committed Choice: The semantics of the language is restricted to the generation of only one solution by choosing a a clause according to its head and some special subgoals in the body, the *guards*, and then committing to that choice – no matter if it turns out to be a good or bad one later on. This restriction avoids backtracking and is sometimes referred to as *don't-care non-determinism* (Shapiro, 1986; Ueda, 1985; Clark and Gregory, 1986).

### Internal Parallelization

Even more than pure OR-parallelism, the combination of AND- and OR-parallelism described here implies a very dynamic execution behavior with a high degree of parallelism. The amount of parallelism is dependent on the number of clause candidates and the number of subgoals of these clauses, and their possible combinations. In practice, this amount of parallelism is way too high to be exploited without some restriction.

---

[19] this means that there is no OR-parallelism for this goal

| AND-Parallelism | | |
| --- | --- | --- |
| | **data structures** | **operations** |
| **computation** | **input data**<br>• formula<br>• (spanning) set of connections<br>**output data**<br>• proof<br>**intermediate data**<br>• AND-parallel tasks<br>• environment<br>• substitutions<br>**evaluation support**<br>• management of variable bindings | **task evaluation**<br>• evaluation mechanisms for tasks<br>**task creation**<br>• creation mechanisms for new tasks<br>**task management**<br>• task distribution,<br>• load balancing<br>• stop evaluation |
| **representation** | **program:** code for<br>• task evaluation<br>• task creation<br>• task management<br>**global data**<br>• formula<br>• proof(s)<br>**local data**<br>• AND-parallel tasks<br>• environment<br>**evaluation support**<br>• variable binding management | **access to code**<br>**access to data**<br>**evaluation support**<br>• management of variable bindings<br>• (distributed) garbage collection<br>• memory management (for multiple tasks on a node) |
| **communication** | **implicit**<br>• shared data structures (shared variables)<br>**explicit**<br>• messages (task creation) | **implicit**<br>• shared data access<br>**explicit**<br>• message passing |

Table 4.19: AND-Parallelism

### Combinations

The combination of AND-parallelism with other categories may lead
to an abundance of parallelism, which can be too difficult or too expensive for practical exploitation. Some categories, however, support
a more structured execution scheme for AND-parallel evaluation of
programs, thus making the management easier.

**AND – Routes**   AND-parallelism and routes present complementary solutions to the same problem, the concurrent evaluation of subgoals; both, however, have their restrictions, and a combination can
be of mutual benefit for both.   Routes can be used to restrict or
structure the abundance of parallelism initiated by AND-parallel execution, whereas AND-parallelism provides the execution support for
those parts of the evaluation which cannot be captured by statical
analysis according to the routes concept.

**AND – OR-Parallelism**   The combination of AND- and OR-
parallelism is a classical form of parallelization for the execution of
PROLOG programs. Although it is conceptually easy to understand,
its realization presents major problems so that in most implementations strong restrictions (e.g. AND-parallelism only for subgoals which
do not share variables) are applied. A promising operational model
for a combination of AND- and OR-parallelism, ANDORRA, has been
proposed by (Haridi and Brand, 1988); it is based on D.H.D. Warren's observation that determinacy should be the basis for a combined
exploitation of AND- and OR-parallelism: goals are executed in AND-
parallel as soon as they become determinate.

**AND – Reductions**   As usual, reductions should be used to reduce the search space before the actual execution of the program. Of
special interest for AND-parallelism are reductions leading to a smaller
number of subgoals in the clauses, or to more independent subgoals
in the sense that shared variables are eliminated where not absolutely
necessary.

**AND – Spanning Sets**   Spanning sets can be used favorably to
structure the evaluation of a formula before the actual execution, thus
decreasing the runtime overhead.

**AND – Competition**   Competition and AND-parallelism can be
combined by using different evaluation mechanisms in a competitive

way for the execution of the single AND-tasks. These different evaluation mechanisms could be based on different calculi, variants of one calculus, or different strategies within one calculus.

**AND – Precision**  A combination of AND-parallelism and precision is to neglect the consistency of shared variables in a superficial proof attempt, but checking it in another more thorough one.

**AND – Modularity**  Although the term 'tasks' has been used in this section to describe the computational units produced by AND-parallelism, according to our terminology for the description of the various categories of parallelism the term 'modules'[20] is more appropriate. Tasks are used for the execution of independent computations, whereas modules feature the execution of computation requiring some exchange of information between the single units.

**AND – Multitasking**  The term 'tasks' nevertheless has a certain justification (apart from the fact that it is common usage in this particular field) in context with AND-parallelism: if AND-parallelism is restricted to the concurrent execution of subgoals without shared variables, tasks are an appropriate mechanism for the execution of the induced computations.

### Conclusions

AND-parallelism is the basis for the execution of the whole family of committed choice[21] languages. In addition, several proposals and also some implementation approaches have been made to combine AND- with OR-parallelism. The amount of parallelism induced thereby can be very large, in many cases too large to be handled reasonably by parallel computer systems.

---

[20] the term 'modules' actually is used with two slightly different meanings in this book; the one used here comprises modules as units of computation to be executed on a computer system; the other views modules as a way of structuring programs (or formulae), concentrating on specification instead of execution
[21] also known as *concurrent logic* and *Guarded Horn Clause languages*

# 4.13  Term Parallelism

The following section is concerned with the exploitation of parallelism
on the *term* level; on that level, the predominant operation is unification, and an alternative name often used is *unification parallelism.*
Since we have to deal with terms, substitutions and unification in
more detail, let us briefly resome the important concepts used (some
have already been introduced in Section 2). Intuitively, terms denote
objects, and unification aims at identifying objects fulfilling the criteria imposed by the terms involved in this unification, resulting in
substitutions or variable bindings for the occurring variables. A *term*
is composed of variable, constant and function symbols. Variables
and constants are terminal symbols and cannot be decomposed further, whereas functions may contain terms as arguments; the number
of arguments a function has is its arity. Terms are typically represented as operator trees or graphs, depicting the relations between
function symbols and their arguments through edges; a somewhat sophisticated variant is to use minimal directed acyclic graphs, or dags,
where multiple occurrences of subdags (esp. variables) are represented
only once.

*Substitutions* assign terms to variables, and the resulting relations
between variables and their assigned terms are called variable bindings. Substitutions are typically denoted by lower-case greek letters
$(\sigma, \tau)$, the application of a substitution $\sigma$ to a term $t$ by $\sigma t$, $\sigma(t)$ or
$t\sigma$; variable bindings are represented as pairs of variables and assigned
terms, with a number of notational variants (e.g. $(v, t)$; $v := t$; $v \leftarrow t$;
$v \setminus t$).

*Unification* of a set of terms $T = \{T_1, \ldots, T_n\}$ is aimed at identifying a substitution $\sigma$ such that the terms after application of this
substitution are identical: $\sigma(T_i) = \sigma(T_j)$ for all $T_i, T_j \in T$.

A method to solve equations which may be considered as the first
unification algorithm has been described in (Herbrand, 1930); a systematic treatment as well as an (exponential) algorithm to compute a
most general unifier has been proposed by (Robinson, 1965). (Martelli
and Montanari, 1982; Paterson and Wegman, 1978) have come up with
more efficient unification algorithms, and some investigations towards
parallel unification have been done by (Yasuura, 1984; Caferra and
Jorrand, 1985; Crammond, 1985; Kung, 1985; Tamaki, 1985; Dwork

**Program TERM**
**parameters**

| *name* | *type* | *initially* | *description* |
|---|---|---|---|
| $dag, dag'$ | directed acyclic graphs | | pair of terms to be unified |
| $root, root'$ | nodes | | starting nodes |

**local variables**

| $FORW$ | set of node pairs | $\{root, root'\}$ | nodes for forwarding |
| $n_i$ | node | | node of the dag |
| $succ_j(n_i)$ | node | | $j$-th successor of node $n_i$ |
| $\sigma$ | variable bindings | $\{\}$ | substitutions |

**invariants**

$meltpair(\,(n_i, n_i'), (n_j, n_j')\,) \equiv (n_i, n_j')$ if $n_i' = n_j$        two pairs of nodes can be melted

$\qquad \wedge\ \exists\,\sigma :: \sigma(n_i) = \sigma(n_i') = \sigma(n_j) = \sigma(n_j')$        if common node and unifiable

$meltset(\{(n_1, \ldots, n_n')\}) \equiv \{i, j :$        melt a set of
$\qquad 1 \le i \le n, 1 \le j \le n ::$        node pairs
$\qquad meltpair(\,(n_i, n_i'), (n_j, n_j')\,)\}$        pairwise

**computation**

$[]\ \langle\ \ i, j : (n_i, n_i') \in FORW ::$        for all node pairs
$\qquad\quad (succ_j(n_i), succ_j(n_i')) ::$        and their successors
$\qquad FORW := melt(\{(succ_j(n_i), succ_j(n_i'))\})$        forward unification and melt 'chains'

$\quad \rangle$

**postconditions**

$FORW = \{\}$        complete dag processed
**end TERM**

Figure 4.20: Program TERM

et al., 1986; Bibel et al., 1987; Harland and Jaffar, 1987; Chen et al., 1988; Citrin, 1988; Delcher and Kasif, 1988; Ramesh et al., 1989; Singhal, 1990). The conclusions from these investigations, however, do not give too much evidence about the suitability of unification for parallelization; some of the results identify a considerable gain of efficiency for restricted problems, whereas others suggest the conclusion that unification inherently is a more sequential process.

Dwork, Kannelakis and Stockmeyer present an algorithm for *term matching* (where a variable may occur only on one side) with a complexity of order $O(\log^2 n)$, $n$ being the number of symbols in the term. Yasuura approaches parallel unification via reachability in directed hypergraphs and shows that it is one of the hardest problems in a class that can be solved by polynomial size circuits. A parallel algorithm based on a *dag* representation for the terms has been outlined in (Bibel et al., 1987); a closer investigation in (Hager and Moser, 1989a) revealed a run-time complexity of $O(\log n)$ in the best, and $O(n/2)$ in the worst case. An implementation of this algorithm on Transputers (Hager and Moser, 1989b) also showed the practicability of the approach for a message-passing architecture, and the suitability for massively parallel systems like a Connection Machine (Hillis, 1985; Thinking Machines, 1987) or a DAP (Hwang and Briggs, 1984) looks quite good. The investigations reported in (Singhal, 1990) show an average speedup of a factor of 2, in some cases 5, over the current highest performance sequential PROLOG implementation. The suitability of different classes of architectures for a parallel implementation of unification based on a dag representation is further investigated in Section 7.4.

In Figure 4.20 a slightly simplified specification of the above algorithm is given; it can be divided into two basic operations. The first is *forwarding:* Starting from the root nodes of the two dags representing the two superterms to be unified, the compatibility of the successor nodes (representing the arguments of the root nodes) is checked, and corresponding pairs of successor nodes serve as roots for subdags to be unified. Such a pair of nodes is referred to as an *arc*, in concordance with its visualization. This process stops as soon as a terminal node occurs in such a pair, resulting in a *terminated arc*. It fails if two nodes of a pair are not compatible (i.e. they have different function or constant symbols). The second basic operation is *chaining:* As an outcome of forwarding, a number of terminated arcs are left, potentially forming a *chain* (if a node appears in two arcs). A consistent binding for all nodes of a chain must be identified now; in the graphical representation this is depicted by 'melting' nodes. This corresponds exactly to the construction of an equivalence class for the nodes involved in a chain. A modification of the *union-find* algorithm (Knuth, 1973) can be used to determine this binding in linear time[22]. Note that in the specification only the forwarding operation is explicitly given; the activities in the chaining phase are achieved implicitly through the invariants for melting sets and pairs of node pairs.

---

[22] this algorithm is described in detail in (Hager and Moser, 1989a)

### Correctness and Complexity

The program TERM specifies the unification of a pair of terms represented as dags; in a recursive way substructures to be unified are identified, and the melting of chains achieves the required substitutions. The complexity of the program is $\sum_{i=1}^{depth}(breadth_i)^2$; the difference to the one described in (Hager and Moser, 1989a) results from the simpler, but less efficient method for melting chains.

### Refinements

An efficient implementation of this algorithm must be based on two basic assumptions: An appropriate representation for the terms to be unified must be chosen, and the size of the terms to be unified must be sufficient for parallel evaluation. The first assumption can be met by the choice of (minimal) *dags*; especially the unique representation of substructures is very helpful during unification[23]. The second assumption concerning the size of the dag does not hold for a common evaluation mechanism where unification is initiated in each evaluation step, and the terms of the single literals are not combined into superterms (as in the spanning set approach). Based on that approach, quite large terms can be obtained (see Section 2.1): The connections of a spanning set represent pairs of literals, whose arguments are terms to be unified. Instead of computing the unifier incrementally by adding the substitutions from one term pair after the other, all term pairs are combined into one huge pair of superterms. This pair of superterms represents all information required for unification in a spanning set, and typically is of sufficient size to be evaluated in parallel. A major problem with the spanning set approach lies in the treatment of recursion; in general, dynamic creation of copies is required, and the substitutions have to be updated. In the worst case this means that the whole pair of superterms has to be traversed again. A suitable internal representation (e.g. hash functions) can reduce this traversal task to reasonable size.

The programs TERM (Figure 4.20) and FORWARD (Figure 4.21) outline the treatment of the categories on the term level. The important features of these categories are briefly described and shown in an overview table.

---

[23] the price for that, however, has to be paid during the construction of the dag (at compile time) where multiple occurrences of substructures have to be mapped onto one instance

**Program FORWARD**

**parameters**

| name | type | initially | description |
|---|---|---|---|
| $dag, dag'$ | directed acyclic graphs | | terms to be unified |
| $root, root'$ | nodes | | starting nodes |
| $n_i, n_i'$ | nodes $\in dag, dag'$ | | nodes the dags |
| $son_{ij}$ | node | | $j$-th son of node $n_i$ |
| $type(n_i')$ | $\in \{\text{nontl}, \text{tl}\}$ | | type of a node |
| $sym(n_i')$ | *string* | | symbol of a node |
| $result$ | $\in \{\text{fwsuc}, \text{sycl}, \text{arcl}\}$ | | possible results |
| $\mathcal{TARCS}$ | set of arcs | | terminated arcs |

**preconditions**

$\exists i : n_i \in dag :: type(n_i) = \text{tl}$          terminal node in the dag

**local variables**

| $\mathcal{ARCS}$ | set of arcs | $(root, root')$ | arcs |
|---|---|---|---|
| $ARC_i$ | node pair | $(n_i, n_i')$ | nodes for forwarding |

**invariants**

$son_{ij} = succ_j(n_i)$          $j$-th successor of node $n_i$

**computation**

$\[\ \langle\ \ i, j :\ ARC_i = (n_i, n_i') \in \mathcal{ARCS},$          for all node pairs
$\quad (son_{ij}, son_{ij}') = (succ_j(n_i), succ_j(n_i')) ::$          and their sons
$\ \|\ \mathcal{ARCS}\ \ := \mathcal{ARCS} \cup (son_{ij}, son_{ij}')$          create new arcs for
$\qquad\qquad\ \ \text{if}\ type(son_{ij}) = \text{nontl}$          two nonterminal nodes
$\qquad\qquad\ \ \ type(son_{ij}') = \text{nontl}$
$\qquad\qquad\ \ \wedge\ sym(son_{ij}) = sym(son_{ij}')$          with the same symbol
$\ \|\ \mathcal{TARCS} := \mathcal{TARCS} \cup (son_{ij}, son_{ij}')$          collect terminated arcs
$\qquad\qquad\ \ \text{if}\ type(son_{ij}) = \text{tl} \vee$          for
$\qquad\qquad\ \ \vee\ type(son_{ij}') = \text{tl}$          terminal node(s)
$\ \|\ result\ \ \ := \text{fwsuc}$          forwarding successful for
$\qquad\qquad\ \ \text{if}\ type(son_{ij}) = type(son_{ij}')$          two non-terminal nodes
$\qquad\qquad\ \ \wedge\ sym(son_{ij}) = sym(son_{ij}')$          with the same symbol
$\qquad\qquad\ \ \wedge\ (son_{ij}) \neq \text{nil}\ \wedge\ (son_{ij}') \neq \text{nil}$ and existing sons
$\ \|\ result\ \ \ := \text{sycl}\ \ \text{if}\ sym(son_{ij}) \neq sym(son_{ij}')$ different symbols
$\ \|\ result\ \ \ := \text{arcl}\ \ \text{if}\ son_{ij} = \text{nil} \vee son_{ij}' = \text{nil}$ different number of sons
$\ \rangle$

**postconditions**

$\mathcal{ARCS} = \{\}$          all arcs processed
$\mathcal{TARCS} \neq \{\}$          some terminated arcs

**end FORWARD**

Figure 4.21: Program FORWARD

## Forwarding

The essential action in the forwarding process is to pass on the information about substructures to be unified. Starting from the root nodes of the dag, the corresponding pairs among the successor nodes are identified and informed that they represent substructures to be unified. The program FORWARD (Figure 4.21) is a refinement and restriction of TERM (Figure 4.20): it only describes the action of forwarding arcs, and delivers as outcome a set of terminated arcs (from which the substitutions have to be computed), and a result about the success or the reasons for failure.

| FORWARDING | | |
|---|---|---|
| | **data structures** | **operations** |
| **compu-<br>tation** | **input data**<br>• term structure<br>• arcs (nodes to be unified)<br>**output data**<br>• terminated arcs<br>**evaluation support**<br>• term scaffold (structure)<br>• tags (types of nodes) | **compatibility check**<br>• same type of nodes,<br>• same number of arguments<br>**mating**<br>• find new pairs of nodes<br>• the new partners are informed about their destiny |
| **repre-<br>sen-<br>tation** | **program**<br>• code to be executed<br>**data**<br>• dag<br>• node pairs representing substructures to be unified<br>**evaluation support**<br>• tags (types of nodes) | **access to code**<br>**access to data**<br>**evaluation support**<br>• support for tags<br>• efficient graph manipulation |
| **com-<br>muni-<br>cation** | **implicit**<br>• shared data structures (dag)<br>**explicit**<br>• messages (arcs) | **implicit**<br>• shared data access<br>**explicit**<br>• message passing |

Table 4.20: Forwarding

## Chaining

The chaining phase represents the actual computation of the substitutions for unification. It works on a set of nodes (transitively) connected by arcs as an outcome of forwarding. The different types and values of these nodes must be checked and made compatible (if possible). It corresponds to the determination of an equivalence class for the nodes, and the most general unifier is the most specific element of that class.

| CHAINING | | |
|---|---|---|
| | **data structures** | **operations** |
| **compu-tation** | **input data**<br>• set of nodes connected by arcs<br>**output data**<br>• variable bindings / equivalence classes | **unification**<br>• determination of equivalence classes |
| **repre-sen-tation** | **program**<br>• determination of equivalence classes, e.g. *union find* algorithm<br>**global data**<br>• set of nodes connected by arcs | **access to code**<br>**access to data** |
| **com-muni-cation** | **implicit**<br>• shared data structures<br>**explicit**<br>• messages | **implicit**<br>• shared data access<br>**explicit**<br>• message passing |

Table 4.21: Chaining

## Occur Check

The operation to perform the occur check is an analysis of the graph for cycles.It can be parallelized in a way similar to the forwarding phase, but starting from the variable nodes and working towards the root. Only combinations of variable nodes and function nodes have to be checked, since that is the only way a recursive substitution can be achieved. *Dags* (directed *acyclic* graphs) initially do not contain cycles[24], but cycles may be introduced in the chaining phase. The occur check must wait until the forwarding and chaining phase have been completed, since otherwise chaining might lead to modifications resulting in cycles not detected by the occur check.

| OCCUR CHECK | | |
|---|---|---|
| | data structures | operations |
| **compu-tation** | **input data**<br>• term graph<br>**output data**<br>• occur check result (truth value)<br>• conflicting parts of the term (clashes)<br>**evaluation support**<br>• term scaffold (structure), e.g. as (connectivity matrix) | **graph search**<br>• check for cycles in the term graph<br>(recursive substitutions) |
| **repre-sen-tation** | **program**<br>• code to be executed<br>**data**<br>• term graph<br>• conflicting parts<br>**evaluation support**<br>• term scaffold (connectivity matrix) | **access to code**<br>**access to data**<br>**evaluation support**<br>• arrays<br>• indexing techniques |
| **com-muni-cation** | **implicit**<br>• shared data structures (dags)<br>**explicit**<br>• messages | **implicit**<br>• shared data access<br>**explicit**<br>• message passing |

Table 4.22: Occur Check

---

[24] although the usage of the term 'dag' is sometimes a little sloppy in the sense that the graphs are not necessarily acyclic by construction, but only after an explicit check

## Bottom-up

In contrast to the forwarding approach, bottom-up starts from the terminal nodes and proceeds towards the root nodes. The information about the substructures to be unified is passed on against the direction of the successor relation. It is especially suitable for programs with many facts or constants, and similar concepts are used in data base systems (Güntzer et al., 1986; Rohmer et al., 1986).

| BOTTOM-UP | | |
|---|---|---|
| | **data structures** | **operations** |
| **compu-tation** | **input data**<br>• term structure<br>• arcs (pair of nodes to be unified)<br>**output data**<br>• substitutions<br>**evaluation support**<br>• term scaffold (structure)<br>• tags (types of nodes) | **compatibility check**<br>• same type of nodes,<br>• same number of arguments<br>**mating**<br>• find new pairs of nodes<br>• inform new partners about their destiny<br>**unification**<br>• computation of substitutions |
| **repre-sen-tation** | **program**<br>• code to be executed<br>**data**<br>• dag<br>• node pairs representing substructures already unified<br>**evaluation support**<br>• tags (types of nodes) | **access to code**<br>**access to data**<br>**evaluation support**<br>• support for tags<br>• efficient graph manipulation |
| **com-muni-cation** | **implicit**<br>• shared data structures (dag)<br>**explicit**<br>• messages (arcs)<br>• substitutions | **implicit**<br>• shared data access<br>**explicit**<br>• message passing |

Table 4.23: Bottom-up

## Separability

Depending on the structure of the formula to be evaluated, the resulting term graph may be separable into several relatively independent subgraphs, which can be processed concurrently. A problem is that an analysis of the overall term structure is required before a separation is possible, although the analysis can be done during the construction of the dag. After the treatment of the single subterms, their variable bindings must be checked for consistency.

| SEPARABILITY | | |
|---|---|---|
| | data structures | operations |
| **compu-tation** | **input data**<br>• pair of (partial) terms<br>**output data**<br>• (partial) unification result (truth value)<br>• (partial) substitutions<br>**evaluation support**<br>• term structure<br>• tags (types of nodes) | **unification**<br>• computation of the substitutions<br>**consistency check**<br>• consistency of the bindings for shared variables |
| **repre-sen-tation** | **program**<br>• code to be executed<br>**data**<br>• pair of (partial) terms<br>• (partial) substitutions<br>**evaluation support**<br>• tags (types of nodes) | **access to code**<br>**access to data**<br>**evaluation support**<br>• support for tags<br>• efficient graph manipulation |
| **com-muni-cation** | **implicit**<br>• shared data structures (dag)<br>**explicit**<br>• messages (arcs, substitutions) | **implicit**<br>• shared data access<br>**explicit**<br>• message passing |

Table 4.24: Separability

## Function Call

Function call stands for the unification and evaluation of function occurrences; it is based on the fact that unification of a (sub-)term with function occurrences can only succeed if the function symbols are identical, occur on corresponding positions, and the evaluations of the functions yield compatible values. Whereas their position can only be checked by a comparison of the term structure 'above', different occurrences of a function can be evaluated before their position is checked. Then the results of the function evaluation are already available, and there is no delay in the unification process. The drawback of this approach is the use of *speculative* parallelism: at the time when the computation is performed, it is not known if the result will be required after all. The benefit may lie in an avoidance of idle times for processing elements.

| FUNCTION CALL | | |
|---|---|---|
| | data structures | operations |
| compu-<br>tation | **input data**<br>• set of function occurrences<br>**output data**<br>• evaluation results<br>• unification result (truth value) | **function evaluation**<br>**unification** |
| repre-<br>sen-<br>tation | **program**<br>• code to be executed<br>**data**<br>• set of function occurrences | **access to code**<br>**access to data** |
| com-<br>muni-<br>cation | **implicit**<br>• function parameters<br>• shared data structures<br>**explicit**<br>• shared data structures<br>• message passing | **implicit**<br>• remote procedure call<br>• shared data access<br>**explicit**<br>• message passing |

Table 4.25: Function Call

## Combinations

The treatment of term structures, especially in the form of unification, occurs in any evaluation mechanism for (predicate) logic, independent of the calculus or strategy applied. Combining term level parallelism with other categories thus is basically always possible, but often practically irrelevant due to the small grain size of the computations involved.

**Term-Parallelism – AND-Parallelism**   AND-parallelism usually is based on an evaluation mechanism which requires the unification of a pair of literals. The sizes of terms involved tend to be too small to be parallelized with good results on architectures with conventional processing elements.

**Term-Parallelism – Routes**   In the routes approach, the literals of linked connections can be melted in one step, resulting in larger term structures than in the usual pairwise fashion. Although a validation in form of an implementation is still missing, concepts of operational models for the routes approach (Ibañez, 1989; Wang, 1989; Balcke, 1987) suggest that a combination with term parallelism is very beneficial.

**Term-Parallelism – OR-Parallelism**   Similar to AND-parallelism, the term sizes in OR-parallel models usually are too small to make use of parallelism on the term level.

**Term-Parallelism – Reductions**   In some reductions, unification is not used, and thus term parallelism is obsolete. In others, the objects investigated are rather small (pairs of literals, unit clauses), and an exploitation of term parallelism is not very promising. The ones dealing with pairs or sets of clauses (subsumption, factorization), are more interesting for an integration of term parallelism.

**Term-Parallelism – Spanning Sets**   A spanning set comprises the connections required for one proof attempt of a particular formula; the validity of the attempt is determined by the unification of all the connections in the spanning set. Since it is possible to tackle this unification task in one big chunk, spanning sets are a very good fit for a combination with term parallelism.

**Term-Parallelism – Competition**   Competition relies on the use of different calculi, or different strategies within one calculus. A combination of term parallelism and competition may be beneficial if the calculus or strategy in use entails the occurrence of large term structures.

**Term-Parallelism – Precision**   Different instances of precision are derived from more or less detailed unification mechanisms. Term parallelism actually provides a parallelization of unification mechanisms, and can be integrated into precision, again provided that the term structures are big enough. Omitting the occur check, for example, can reduce the time required for unification roughly by a factor of two: the occur check can only be performed after the forwarding and chaining phases, and its complexity is about the same as the one of the two other phases together.

**Term-Parallelism – Modularity**   Although a separation of a formula into modules decreases the potential size of the resulting term structures, it provides a way of structuring which can lead to a better applicability of concepts like spanning sets and routes, which in turn highly favor the exploitation of term parallelism.

**Term-Parallelism – Multitasking**   The grain size induced by a term parallelism is too small for typical multitasking systems; if multitasking (or something similar)[25] can be used on a massively parallel machine, such a system provides an excellent platform for the combination of term parallelism and multitasking.

### Conclusions

In principle, parallelism on the term level can be exploited whenever terms have to be unified, independent of the particular evaluation mechanism. In practice the computation units resulting from parallelization on the term level are very small, and the administrative overhead can easily exceeds the potential gain of performance. But based on processing elements with fast task switching, and term sizes of dozens of symbols, a practically relevant speedup can be obtained (Singhal, 1990). Taking into account that the operation of conventional processing elements is not designed for unification as basic operation, the potential of parallelism on the term level is quite good.

---

[25] for a more detailed discussion, a separation between *temporal* multitasking (time-sharing) and *spatial* multitasking (space-sharing) would be necessary

# 4.14   Distributed Representation

This section deals with an alternative paradigm to represent data in an information processing system. In conventional systems a *local representation*, or one-to-one mapping, can be found between atomic objects and basic representation units (e.g. variable or constant to memory cell); in the paradigm investigated here the representation of an atomic object is *distributed* over a number of units, and each unit in turn contains fractions of more than one object (Hinton, 1984; Rumelhart et al., 1986; Palm, 1980). Based on the distributed representation paradigm, a whole variety of computational models can be designed for processing the represented information.

In the context of this book, the main interest lies in an identification of potential parallelism applied to the evaluation of logic formulae. Since it is not feasible here to discuss all the different models and their relevance for logic and parallelism, some essential features of neural models are presented; their application is outlined as an evaluation mechanism for classical logic, and as a basis for reasoning tools not necessarily derived from formal logic. Due to these reasons, the structure of this section deviates somewhat from the typical scheme used throughout the rest of the chapter; instead of presenting one particular problem with a solution and a formal specification, more general aspects are described and discussed with respect to their relevance for parallelism in logic.


## Neural Models

Functional models of neurons provide a methodology for processing information which is quite different from conventional numerical or symbolic computation. It seems to offer some advantages for the execution of tasks where an algorithmic or formal solution cannot be found easily, or is not appropriate. In many cases, however, the structure of the problem to be solved must be represented in the structure of the neural model – and this task still must be done by the designer of the model (Feldman et al., 1988). In addition, investigations of the underlying theoretical basis as well as experiences with practical applications are still in their initial phases. As a consequence, well-founded statements about properties of such a model are rather difficult to obtain.

Neural models typically comprise a collection of nodes with the following features:

- a huge number of simple elements (nodes),
- a node has a functionality similar to a neuron,
- a dense interconnection scheme,
- a highly parallel mode of operation,
- no (explicit) programming.

Customizing a neural model for the solution of a specific task can be done either by a description of its internal architecture (*structured* model), or by learning through the presentation of some (hundreds or thousands of) training patterns (*unstructured* model).

The behavior of a neural model depends on a variety of parameters:

**Static features**
- sets of possible *input* values, *output* values, internal *states* and transitions, and *interconnection* strengths;
- network *size* (number of nodes);
- network *topology* with the interconnections between nodes.

**Dynamic features**
- input function: Computes the actual input into a node from the values arriving via input connections;
- transition (activation) function: Computes the new internal state from the input value, the old internal state, and (possibly) the old output value;
- output function: Computes the values to be fed into the output connections from the output value (or the internal state) of a node;
- transmission function: Computes the change of a value during its transmission.

Adapting a neural model to a task (or class of tasks) requires an appropriate setting of these characteristics. Two fundamentally different but also intertwined aspects, *representation* and *processing* of information are described in more detail below.

## Representation of Information

Neural models offer a storage and retrieval medium with some advantages over conventional computer memories. First, they provide content-oriented access to the information stored (associative memory); second, they show some degree of internal self-organization;

third, they are much more tolerant towards faults or noise in the input
data during retrieval as well as internal corruption. Disadvantages lie
in a higher access time (compared with raw access to semiconductor
memory), and a lower storage density[26]. The crucial point here is
*storage*, not processing of data items.

**Structured models:** Some essential (meta-) knowledge about the
information to be represented is already given implicitly by the designer (interconnection scheme, evaluation functions, allocation).

**Unstructured models:** the structure of the internal representation
is derived from the input values and the items already stored; this requires some kind of (meta-) mechanism to evolve the representation
characteristics (self-organization, abstraction, learning).

## Information Processing

Very often the task to be accomplished by a neural model involves not
only storing and retrieving information, but *deriving new information*
from already known data. To some degree this is already done during
the representation process of unstructured models (learning phase). In
the following we briefly characterize some basic information processing
tasks

- Comparison of structures, e.g. identity or similarity (pattern matching), inclusion (searching for subgraphs in a graph).
- Abstraction and generalization by extracting common features from
  a set of items.
- Completion by adding specific features to an incomplete set of features.
- Search for distinguished states, e.g. global or local maxima / minima
  (esp. energy minimization).
- Unification by finding an instantiation acceptable for all partners
  (structures) involved.
- Feature determination by extracting information from the structures
  under evaluation (number of elements, diameter, shortest path).

Many of these basic information processing tasks seem to be quite
useful for the incorporation into a logic programming system, either
as basic primitives (e.g. unification), or for adding valuable support
to conventional logic systems. In principle it should also be possible

---

[26] in certain models, however, the utilization ratio can be around 70% (Palm,
1980).

to build a self-contained reasoning system based on neural modeling methodologies.

## Reasoning with Neural Models

The momentary fascination with neural models emanates from at least two sources. One surely is the relationship to what's going on in our brain (or at least what we think is going on there), and the other is the understanding that conventional computers have certain flaws[27], especially with tasks that seem to be almost trivial for natural information processing systems.

The ability to draw conclusions from statements or to check out the truth of such conclusions surely is one of the facets which contribute to intelligent thinking. The method of deriving new information from already known items to some degree is based on

- *"logical" rules* as abstract formalism[28] for drawing conclusions, and
- *neurons* as physical computation units.

Of course intelligence and human thinking comprise much more than is mentioned here, and maybe it is somehow vain to try to mimic one particular feature without taking into account all the other ones. But still experimentation based on artificial models which at least have a few commonalities with the real thing might be useful for gaining some insight into the way human beings use their brains. The question arising here is if it is possible to synthesize a model which follows the same or equivalent "logical" rules and is based on the same way (paradigm) of computation as human reasoning.

Checking out the substantial truth value of a chain of conclusions, or more general, a set of statements, sometimes can be quite a non-trivial task, maybe even more so if it is not done within a formal and safe framework such as mathematical logic[29]. Quite a helpful tool for fulfilling such a task might be something like a "truth checker", which tries to find the "maximal logical consistency" of a collection of statements by checking out all (or at least most) of the possible combinations, and choosing one which violates only a few (or none in the optimal case) conditions. In terms of neural models, this method-ology is quite close to *energy minimization*, where a physical system

---

[27] despite all the efforts to stuff them with artificial intelligence

[28] this abstract formalism usually is not explicitly given, and very often is subject to manipulation, if not fraud

[29] for a typical instance see the area of laws and jurisdiction

is modeled in order to find the state with the lowest (or a sufficiently low) energy level.

A possible outline for modeling such a system looking for maximal logical consistency could be based on a graph representation of the logical program (similar to a connection graph in formal logic), which represents the interrelations between the single statements, and "melt" corresponding nodes (statements) together (sloppy unification), if they are sufficiently consistent. The result here would be a huge chunk of melted statements, and it might be difficult to extract any useful information from it. An alternative is to arrange a set of $n$ atomic statements into a $n$-dimensional space, connect them according to their interrelations, and try to minimize the "tension" in the system by varying the degree of "satisfaction" of the single statements; differing importance of statements could be represented by stronger springs in the nodes, thus resisting dissatisfaction more vigorously.

A system capable of drawing conclusions, not only checking them, has to face another, fundamentally quite different problem. It has to be able to choose a favorable one from a set of possible conclusions. This step is the one where conventional, symbolic reasoning systems probably have the biggest difficulties; it is not easy to capture this step algorithmically, and – when applied by humans – involves experience, intuition, creativity, phantasy — concepts that are far beyond the scope of conventional computer systems.

## Logic with Neural Models

At a first glance, the combination of neural models (with their some-what sloppy evaluation paradigm) and mathematical logic (with its strongly formal and exact aspects) seems a bit awkward, maybe like hammering a nail into a concrete wall with a rubber tool. A second thought, however, reveals that neural models have some strength where formal logic has its problems: the description of "real world problems" with inconsistencies, exceptions to the rules, incomplete specifications, uncertainties, changes in the course of time, and similar characteristics. A lot of effort has been invested in the adaptation of formal logic towards capturing these features (e.g. modal logic, temporal logic, non-monotonic reasoning, uncertain reasoning, truth-maintenance systems), but most of the approaches are still too restricted, too complicated to be dealt with efficiently on conventional computers, too unnatural for adequate specification, or all of the above.

Storage and processing of information with neural networks is more tolerant towards inconsistencies, incomplete information, and temporal changes. Their use in combination with formal logic might help to some degree in overcoming the problems mentioned above; however, some of the strong formal properties of logic (correctness and completeness) might be affected in an negative way.

**Propositional Logic** The treatment of propositional logic is obviously much easier than dealing immediately with predicate logic. In the following we discuss some possibilities for the use of neural models in propositional logic.

1. Formula evaluation, which consists of a computation of the output value from given input values; the formula can be in normal form, or non-normal form, where additional treatment of the formula structure is required. The basic task here is to construct a network that corresponds to the formula under investigation (Neumayer, 1988); it is closely related to Boolean logic and the composition of digital circuits from basic logic gates.

2. Satisfiability, or finding out if there is an interpretation for the formula such that the formula is true; in this case, many sets of input values (which correspond to the interpretations) might have to be checked out. This problem can be solved by mapping it onto an equivalent one, which then is treated by a neural model. Two candidates for such an equivalent problem are graph coloring and gradient method (Mayr and Reich, 1988; Kurfeß and Reich, 1989).

3. Validity, where the formula is true for every interpretation. Here all possible interpretations have to be checked out, either by a "brute force" or more sophisticated methods, e.g. abstraction of the previous models, or evaluation of the connection graph.

**Predicate Logic** The treatment of predicate logic has to face some additional problems, like dealing with logical variables and their bindings (and thus unification), function evaluation, and clause copies for recursive problems. These problems are quite difficult to solve with neural models, and in some cases straightforward solutions are not immediately visible. However, not much work has been done yet in this direction, so that there is still some hope left for satisfying answers, or at least deeper understanding.

Horn clause logic can be tackled by

- reproducing the rules of a calculus, e.g. resolution (Ballard, 1986), or

- evaluation of the connection graph (including unification) (Hölldobler, 1990a).

Full first order predicate logic may be treated

- using *matrix logic* (Stern, 1988), interpreting logical connectives as matrix operators acting on the two adjoint vector spaces (logic as a "macroanalogue of spin");

- via "lifting" the graph coloring approach for propositional logic (graph nodes represent terms instead of simple symbols);

- via "lifting" the gradient method for propositional logic (higher order functions);

- by an evaluation of the connection graph.

Maybe the most promising approach for the time being consists of using neural network methodologies for the accomplishment of tasks which cannot be treated very well with conventional (algorithmical, symbolical) techniques. It has the advantage that existing systems, together with the knowledge acquired during their construction, can be used with minor modifications and enhancements.

The evaluation of logic programs offers quite a few possibilities for integrating services to be yielded by neural models.

**Choice of Strategies and Heuristics:** A neural network is used to select an appropriate strategy for a particular formula:

- *Global strategies*, where a choice strategy is made once at the beginning of the proof and thus determines the overall proof process. Examples are
    - o scope: propositional or predicate logic;
    - o calculus: resolution, model elimination / connection method, natural deduction;
    - o search strategy: depth- / breadth-first;
    - o depth boundary (for depth-first)
      (Schumann et al., 1989; Suttner, 1989).

- *Local strategies*, where decisions are made according to the current, local proof environment and are applied only during a limited period. Examples are o computation strategy: execution order of the subgoals (rearrangement of the goal stack);

o local search strategy: choice of branches (Schumann et al., 1989);

o choice of an appropriate unification algorithm (full / partial; different versions)

o creation and application of lemmata;

o "intelligent" backtracking (Bruynooghe and Pereira, 1984).

The choice of a strategy heavily depends on the features taken into account. In the case of neural networks, this *feature extraction* can be provided by training examples, by previously solved tasks, or by the system designer.

**Reductions:** The successful application of reductions to a formula depends on its actual structure. Some of the reduction steps are quite powerful (and may even lead to a proof or refutation of the formula), but also computationally expensive; thus it is advisable to first check out whether it is worthwhile to evoke a particular reduction step. Again in this case a neural model might be more appropriate since the task of choosing the appropriate reduction is difficult to capture algorithmically.

• *Global reductions*, where an application of the reduction requires information about the whole formula. Examples are pure literals or isolated connections.

• *Local reductions*, where it is sufficient to take into account only a part of the formula (single clauses, pairs of clauses). Examples are tautology reduction (single clause), restricted unit resolution (pair of clauses), subsumption (pair of clauses), or factorization (single clause).

• *Combinations of reductions* can lead to mutual enforcement (snowball effect), resulting in drastically smaller (in the extreme case empty) search spaces.

**Matching and Unification:** Matching in the simplest case consists of the comparison of two structures (patterns, graphs, terms) for identity. Unification is more complicated since variables can be involved which have to be substituted during the unification process.

• *Pattern Matching* involves no variables.

• *Term Matching (One-way Unification)* has to deal with variables on only one side at a time.

• *Restricted Unification* is constrained by annotations indicating the direction of the binding process (e.g read-only variables).

- *Full Unification* of a pair of terms, without restrictions.
- *Set Unification*, where many terms are unified in a single run.
  Approaches to unification with neural models are described in (Stolcke, 1988; Hölldobler, 1990b; Kurfeß, 1991).

## Related Work

Investigations concerning the combination of neural models (neural networks, connectionism, parallel distributed processing) with logic and reasoning are still relatively few; some approaches[30] are

- a logical calculus of the ideas immanent in nervous activity (McCulloch and Pitts, 1943),
- a problem-solving computer, based on access to information according to classification rather than addresses (Widrow and Hoff, 1960),
- a formal theory of evidential reasoning in neural networks (Shastri and Feldman, 1985; Shastri, 1988; Shastri and Ajjanagadde, 1989; Shastri and Ajjanagadde, 1990),
- a production system interpreter, where a successful rule match corresponds to a low energy state (Touretzky and Hinton, 1985),
- a resolution inference mechanism based on energy minimization (Ballard, 1986),
- an integration of neural models into expert systems (Giambiasi et al., 1989),
- an inference system for Horn clauses (with some restrictions) based on the connection method using connectionist techniques (Hölldobler, 1990a).

Interest in the connection of symbolic computation and neural models is growing rapidly, and many exciting and surprising, but maybe also frustrating, results are to be expected.


## Combinations

The main potential of distributed representation paradigms lies in additional support for the evaluation of logic formulae with conventional models. They are particularly useful in situations where the extraction of information via algorithmic or symbolic methods is difficult.

**Distributed Representations — Term Parallelism**    Two possible applications of neural models, and also distributed representation

---

[30] often only briefly mentioned, however

| DISTRIBUTED REPRESENTATION | | |
|---|---|---|
| | **data structures** | **operations** |
| **computation** | **input data**<br>• formula<br>**output data**<br>• proof result (truth value)<br>• substitutions<br>**internal data**<br>• state of the network<br>**evaluation support**<br>• previously learned information<br>• structure of the network | **proof** (recall phase)<br>• state transitions according to the activation function<br>• structure of the network<br>**learning phase**<br>• modifications of the network structure |
| **representation** | **program**<br>• input function<br>• output function<br>• activation function<br>• transmission function<br>**data**<br>• formula<br>• substitutions<br>**internal data**<br>• state of the network<br>• network size<br>• network topology<br>**evaluation support**<br>• previously learned information<br>• structure of the network | **structure**<br>• modifications of the network structure<br>**access**<br>• current values (state) |
| **communication** | **implicit**<br>• distributed representation (weight coefficients of links between nodes)<br>**explicit**<br>• values of input and output functions | **implicit**<br>• adjusting the weight of links<br>**explicit**<br>• exchange of values according to the transmission function |

Table 4.26: Distributed Representation

in general, are unification as a whole, and the occur check. A general problem with the combination of distributed representations and unification is the treatment of variables, which requires some additional measures, e.g. intermediate representations to preserve the information where the variable bindings come from (Dolan, 1989), or explicit binding cells (Touretzky and Hinton, 1985). An approach to unification with neural models has been described in (Stolcke, 1988). The occur check detects cycles in a graph; this task can be done using a neural model which has been trained by presenting graphs with and without cycles.

**Distributed Representations — AND-Parallelism**   Neural models can support AND-parallel execution of logic programs by providing information about the selection of subgoals, e.g. parallel or sequential evaluation of a particular subgoal, or selecting 'promising' subgoals to be handled first.

**Distributed Representations — Routes**   Applications of neural models to routes are their detection in a connection graph, and a classification for further treatment (e.g. the need for copies).

**Distributed Representations — OR-Parallelism**   Similar to AND-parallelism, neural models can provide information for the application of heuristics at certain points in the execution process. For OR-parallelism this can be the selection of clauses for parallel treatment, as well as decisions about the sequence in which the clauses are used (provided that parallelism must be restricted).

**Distributed Representations — Spanning Sets**   Spanning sets can profit from an application of neural models in three aspects: first, the generation of sets can be guided towards constructing sets which very likely are spanning; second, the spanning property can be checked by a neural network, and third, the validation of a spanning set (its unifiability) can be checked by neural models.

**Distributed Representations — Reductions**   Some reductions can be viewed as detection of patterns in a representation of the formula; this task can either be directly done by neural models, or at least supported, e.g. by selecting parts of the clause which are good candidates for a certain reduction step, or by checking if parts of the formula may be reduced successfully.

**Distributed Representations — Competition** An evaluation mechanism based on neural models can be one of a set of competitive provers; furthermore, neural models may be used to select appropriate evaluation mechanisms for particular formulae.

**Distributed Representations — Precision** Neural models can provide the basis for evaluation mechanisms which are not fully correct from the logical point of view, but detect essential properties of (classes of) logical formula with good reliability.

**Distributed Representations — Modularity** Applications for neural models to modularity comprise support for structuring formulae into modules, or for the execution of modules on a particular architecture (e.g. allocation, load balancing).

**Distributed Representations — Multitasking** Similar to modularity, neural models can provide support for the execution of multiple tasks on a parallel computer system.

## Conclusions

The distributed representation paradigm in general, and neural models in particular, can be used to provide support for conventional logic evaluation mechanism, to implement such mechanisms, or to construct reasoning systems closer to human thinking than formal logic. Their practical usability is difficult to estimate since research in that field is still in its initial stages; experimentations have shown that there are applications where the evaluation of logic programs can profit from support through neural models (Suttner, 1989; Schumann et al., 1989), and that neural models can be used as evaluation mechanisms for restricted classes of logical problems (Mayr and Reich, 1988; Kurfeß and Reich, 1989; Hölldobler, 1990b; Hölldobler, 1990a; Kurfeß, 1991).

# 4.15   Conclusions: Parallelism in Logic

In this chapter it could be shown that there is a whole variety of possibilities to exploit parallelism in logic, and that some of the categories can be combined with mutual benefits. The main sources of parallelism are the reflection of the *problem structure* to be described in the logical formula, and particular properties of the *evaluation mechanism* used. The first one mainly depends on the user and the problems envisaged, but language construct must be provided to adequately describe these problems. Modularity (or related concepts like objects) represent a good basis for a structured problem description.

An important criterion for a parallelization of the evaluation mechanism is the availability of independent computational tasks at evaluation time. A good candidate should not rely on global data structures, allow the identification of parallelism at compile time, and avoid complications at run time (e.g. context switches, load balancing, distributed memory management).

Most attempts to parallelize PROLOG do not come off very well with respect to these criteria; in addition, they either are restricted by sticking to the original sequential semantics, or must give up compatibility.

An evaluation mechanism with better features regarding parallelism will be presented in the following chapters. It is based on the spanning set and routes concepts, and relies on a statical analysis to identify alternative solutions (spanning sets / OR-parallelism) and sequences of connections (routes). As a result, data are only used locally, and run-time management is reduced as far as possible. Furthermore, the treatment of the term structures can be separated from the handling of the logical relations, leading to larger terms to be unified and better usage of parallelism at the term level. It is open towards a straightforward integration of additional categories like reductions, competition, and precision, as well as modularization and meta-evaluation. Finally, a clear declarative semantics is maintained whilst allowing user control of the execution, e.g. for efficiency reasons or for experimentation.

# Chapter 5

# A Parallel Logic Language: MMLOP

In this chapter, a language based on logic is sketched which aims at an expression of the various categories of parallelism described before. Corresponding to two major sources of parallelism, *problem-induced* and *execution-oriented*, this language offers two classes of constructs to express parallelism: *modules* and *meta-evaluation*. Modules aim at an appropriate description of problems which can be decomposed into relatively independent units of computation; meta-evaluation allows for a proper expression of knowledge about the evaluation and execution of logic programs. The combination of these two basic concepts and their integration into the first order logic language LOP leads to a framework for parallel programming suitable for high-level specification and low-level coding of parallel problems. It unifies the declarative character of *logic programming* with fundamental concepts necessary for *parallel programming*.

## 5.1  Overview

Many languages for parallel programming suffer from a missing separation of *problem-induced* and *execution-oriented* parallelism. The term *problem-induced* is used to denote parallelism which mirrors the structure of the problem to be solved and is largely independent of the way the computations are executed to obtain a solution, whereas *execution-oriented* parallelism is mainly influenced by the computation process itself and its organization according to some evaluation mechanism on a specific parallel architecture. These two terms do not necessarily coincide with the notions of implicit and explicit parallelism: the latter are used to denote how parallelism is obtained and controlled, whereas problem-induced and execution-oriented express the origin of parallelism.

Some languages offer appropriate constructs to describe more or less independent units of computation (e.g. via modules, objects, processes), but show weaknesses in control information for the exploitation of parallelism inherent in the evaluation mechanism, whereas others require the use of evaluation control constructs to model problem-induced parallelism (e.g. guards, annotated variables, messages). In favor of the first class it is often argued that evaluation control on the user level is not necessary or unwanted, and should be left to an intelligent compiler or evaluation support tool. This argument is based on the very desirable, but at least for the near future unrealistic availability of such intelligent tools. The second class with direct control of the evaluation through the user surely has its appropriate application domains, but is less suitable for high-level specification of parallel systems independent of the underlying language evaluation mechanism.

In this section, requirements on a parallel programming language are derived using the categories of parallelism in logic described in the previous chapter; *language constructs* together with *control facilities* are identified to express the phenomena underlying the different categories of parallelism, and possible *computation mechanisms* are discussed to evaluate these language constructs. The discussion is carried out in three parts, proceeding from *coarse* over *medium* to *fine grain*; important aspects are summarized in overview tables.

## Coarse Grain

Categories mainly concerning phenomena dealing with relatively large units of computation are multitasking, modularity, competition, precision, spanning sets, and some reductions. Their scope covers whole formulae or large parts of a formula, structuring the problem to be treated or describing different ways to evaluate the formula. The expression of these features in an appropriate way requires constructs which go beyond the scope of first order predicate logic. This can be done in an ad hoc way by introducing additional constructs like `begin / end module` (together with `import / export`), which can be eliminated via renaming of predicates in the symbol tables of a compiler, for example; that's the way modules are realized in many PROLOG systems, e.g. PEPSYS (Rapp, 1988). A more sophisticated way is to use proper second order logic constructs, e.g. second order predicate symbols. A few concepts based on a higher order logic approach have been proposed (for an overview see (Barbuti et al., 1987)), although mainly directed towards integrating concepts like procedure definition / call mechanisms, hierarchization and modularization, or knowledge representation aspects like managing dynamically evolving knowledge. Similar aims have been pursued by adding or integrating object-oriented concepts to PROLOG (Shapiro, 1986; Tokoro and Ishikawa, 1986).

### Modules

A proper definition of *modules* using second order predicates (where module names are second order predicates) and an *algebra of modules* to define their relations seems to be a well-founded and appropriate way to express parallelism with a relatively large grain size in a logic programming framework. In addition to a natural description of parallel problems, modules can be used to provide higher-level programming constructs like abstract data types (genericity) as well as libraries containing modules of general interest, or dedicated modules for specialized applications. The concept of modules can be supported statically (at compile time), or dynamically (at run time). The static case is relatively easy to implement (and actually present in some PROLOG systems), but has some obvious limitations; the dynamic case is more difficult to implement, but offers possibilities for (re-)arranging modules at run time (e.g. for load balancing).

A module typically comprises a relatively independent computational task; nevertheless, information has to be exchanged between modules forming a more complex system, and both for the descriptive power and an efficient evaluation an appropriate communication mechanism. Ways to express communication between modules are

- via *quantification* of (second order predicate) variables: universally quantified variables are imported, existentially quantified ones exported;
- implicitly through their occurrence in the module definition: exported predicates occur in the argument list of the module definition and in the clause headers, imported predicates in the argument list and the clause bodies, and internal predicate occur only inside a definition of a module.

This concept of modules allows for a straightforward way of structuring programs; it is also suitable for multiprogramming and multitasking, where different programs are represented by modules without common predicates. The evaluation of logic programs based on modules can be achieved with the following computation mechanisms:

- procedures: modules as procedure definitions;
- processes: each module is executed by a process, or a set of processes;
- objects: each module corresponds to an object;
- abstract data types: a module defines an abstract data type;
- partial evaluation: (generic) modules are instantiated and specialized as far as possible;
- algebra of programs: a set of modules and their inter-relations are interpreted as an algebra of programs with corresponding composition operators.

These evaluation mechanisms can be combined to yield a flexible, powerful, and also efficient execution of modular logic programs: Partial evaluation transforms a set of generic modules, connected by composition operators, into specialized ones; these are mapped onto corresponding objects with their access mechanisms, and the objects are executed by a network of communicating processes.

### Theories and Meta-Programming

An extension of the module concept can also be used to handle features sometimes called *meta-programming*, or *worlds* in artificial intelligence terminology; a common term in a logic environment is *theories*. The main goal here is not a structuring of the program, but an adaptation of the evaluation mechanism by an amalgamation of (object-) language and meta-language. A number of similar approaches have been proposed (Bowen and Kowalski, 1982; Lloyd, 1989), most of them aiming at higher-order inference mechanisms without specific relation to parallelism. A theory is associated to a specific inference mechanism, together with operations to use and modify the theory as well as (meta-)program development tools. Theories can be connected by (hierarchical) links to express inheritance (with the meta-predicates is_a, part_of). With respect to the categories of parallelism defined previously, theories are appropriate to describe

- competition: theories representing different calculi or strategies to evaluate a program,
- precision: theories based on unification mechanisms with different degrees of accuracy,
- spanning sets: theories with statical identification of alternative solutions,
- reductions: theories performing transformations of the (object) program to trim down the search space.

Candidates for computation mechanisms to evaluate programs using theories are:

- meta-programming: definition on top of existing (object level) languages, like PROLOG or LOP,
- partial evaluation: specializations of meta-programs are generated, and
- abstract interpretation: a program is interpreted without input parameters (query), thus reducing the code to be executed later.

The combination of meta-interpreters and partial evaluation offers a program transformation and compilation technique which has a sound logical background, is very powerful and flexible, and can also be quite efficient since it results in directly executable object programs. Basic constructs for meta-programming are usually provided in PROLOG systems; typically they are used for rapid prototyping purposes or adaptations to specific problem areas in form of meta-interpreters (Shapiro, 1986; Tanaka, 1988; Silver, 1986; Takeuchi and Furukawa, 1986). In addition to that, meta-programming has the potential to increase the efficiency of logic program execution substantially, especially in combination with sophisticated compilation techniques such as partial evaluation, and with parallel architectures as execution vehicles (Goto et al., 1988).

| *category* | *language construct* | *implementation* |
|---|---|---|
| • multitasking (multiprogramming, multiuser), <br> • modularity | • modules: `begin / end module, import / export`, <br> • second order predicates, <br> • abstract data types, <br> • objects | • partial evaluation, <br> • algebra of programs, <br> • objects, <br> • procedure definition / call, <br> • renaming of predicates (symbol tables), <br> • macro expansion |
| • calculi, <br> • strategies, <br> • precision, <br> • spanning sets | • theories, <br> • worlds, <br> • meta-level constructs, <br> • frames | • theory processors (specialized inference mechanisms), <br> • meta-interpreters, <br> • partial evaluation, <br> • algebra of programs, <br> • objects |

Table 5.1: Parallel phenomena and their description: Coarse grain

## Medium Grain

Categories and phenomena of parallelism in logic programming dealing with medium size computational units are *OR-parallelism* (internal or local non-determinism), *AND-parallelism*, external or global non-determinism, communication and synchronization, the usage of *intermediate results*, and the treatment of *large data structures*.

Language constructs to describe these phenomena are

- alternative clauses for OR-parallelism; different solutions for a subgoal can be achieved if there are more than one clauses whose head matches the subgoal.

- alternatives in the clause body for external non-determinism; within one and the same clause, alternative definitions for its body can be given by providing alternative sets of subgoals (Cunha et al., 1989). The choice of the alternative depends on values external to the formula, determined by the environment.

- multiple subgoals for AND-parallelism; if the sequence in which the subgoals of a clause body are evaluated is of no importance, these subgoals can be treated in parallel. This can be either implicitly assumed in the computational model (which results in a different semantics)[1], or explicitly expressed by separate constructs for sequential and parallel conjunction of subgoals. Consistency of common variables in parallel subgoals has to be maintained, e.g. by an additional unification after the evaluation of the subgoals.

- events for communication and synchronization; in addition to the exchange of information through logical variables, a communication mechanism based on term unification can be provided by special predicates. Various annotations can be used to express directed or undirected and synchronous or asynchronous communication. Attention has to be paid to the underlying computational model, which may affect the semantics of unification. In addition, the backtracking mechanism may have to be enhanced to distributed backtracking in the case that an event goal fails, or is involved in (local) backtracking. This problem can be diminished by the introduction of separate constructs for backtrackable and non-backtrackable events; in the latter case, however, the search strategy becomes incomplete in the case that both event goals involved fail.

- lemmata for intermediate results; avoiding repeated computation of intermediate results can enhance efficiency considerably; these intermediate results can be represented in the form of lemmata, whose generation can be controlled implicitly through the underlying computation mechanism, or explicitly by marking promising occurrences.

Handling of large data structures usually is not dealt with on the language construct level, but more in the computation mechanisms described below.

A computation mechanism for AND-/OR-parallelism is based on a typical evaluation mechanism for logic programs, e.g. resolution for PROLOG. The semantics of this mechanism can either be maintained to include the sequential part as a proper subset and enhanced by compatible parallel constructs (as in "all-solutions languages" like ANDORRA, Delta PROLOG, AURORA, PEPSYS, ...), or it can be adapted to parallel evaluation without maintaining compatibility (as

---

[1] as in *committed choice languages* like Concurrent PROLOG, Parlog, and GHC

| *category* | *language construct* | *implementation* |
|---|---|---|
| • reductions | • implicit in the computation mechanism,<br>• meta-programming | • syntactic transformations |
| • OR-parallelism (internal non-determinism) | • alternative clauses | • derivation tree (AND/OR–tree),<br>• hierarchy of processes,<br>• objects;<br>• backtracking,<br>• reproof mode,<br>• binding list |
| • external non-determinism | • alternatives in the clause body | • selection according to values delivered from the environment |
| • AND-parallelism | • multiple goals: parallel composition operators (explicit),<br>• modified semantics (implicit) | • derivation tree,<br>• processes,<br>• objects;<br>• consistency check (shared variables),<br>• distributed backtracking |
| • communication,<br>• synchronization | • events (special predicates) | • unification,<br>• distributed backtracking;<br>• message passing,<br>• shared memory |
| • intermediate results | • lemmata | • additional clauses |
| • large data structures | • abstract data types,<br>• objects | • objects;<br>• lazy evaluation,<br>• partial computation |
| • sequences | • lists,<br>• sets | • streams,<br>• partial computation |

Table 5.2: Parallel phenomena and their description: Medium grain

in "committed-choice languages" like Concurrent PROLOG, Parlog, and GHC). A computation mechanism for communication and synchronization via events is a "rendez-vous" concept based on term unification, which is a generalization of the synchronous communication between processes described in (Milner, 1980; Hoare, 1985). Basically the same concept is used in the functional parallel language FP2 (Jorrand, 1986; Schnoebelen and Jorrand, 1989). The case for lemmata is quite simple: they just are added to the original program as extra clauses. Handling of large data structures can be achieved through lazy evaluation or partial computation, a particular instance being streams for long sequences of data.

## Fine Grain

Small units of computation are found in term unification (forwarding, bottom up, chaining, function evaluation), and sub-symbolic models based on distributed representation. The language constructs for their description are normally present in a logical language anyway; terms appear as arguments of predicates, functions are integrated in terms, and chaining is given by sets of complementary literals; sub-symbolic computation is not apparent in the language itself. The underlying computation mechanisms for the first three are quite conventional. For the last, connectionist models can be applied; a potential usage lies in a "fuzzy" unification mechanism able to deal with inconsistent or incomplete information.

Typical parallel languages based on logic tackle the use of parallelism on the *medium grain* level, e.g. Concurrent PROLOG, PARLOG, GHC, Delta PROLOG, ANDORRA, PEPSYS, and others. This might be useful for applications dealing with algorithms derived from, or reasonably close to sequential solutions (which in fact is a basic requirement for some languages because of their maintenance of the standard PROLOG semantics). For the description of systems composed of many, largely independent functional units with clearly identified interconnections, these approaches have some severe limitations, mainly due to their lack of well-founded and clean constructs to describe *coarse grain* phenomena.

The proposal here is to design a parallel logic language based on *modules* and *meta-evaluation*. Modules are used to describe systems

| *category* | *language construct* | *implementation* |
|---|---|---|
| • chaining | • implicit: sets of complementary literals | • unification of sets of terms,<br>• determination of equivalence classes |
| • term unification | • implicit: terms as arguments of predicates | • nested lists,<br>• superterms (e.g. for modules),<br>• dags |
| • function evaluation | • implicit: function symbols + parameters | • procedures,<br>• term / graph reduction |
| • sub-symbolic | • implicit: symbols | • connectionist models,<br>• statistical methods (e.g. Bayesian inference) |

Table 5.3: Parallel phenomena and their description: Fine grain

composed of largely independent units of computation in a natural way; metaprogramming techniques (or theories) are applied for the structuring of systems into modules, the connections between the modules, and the evaluation of programs specifying single modules. An integration of metaprogramming constructs (independent of modularity) into LOP is outlined in (Kurfeß and Schumann, 1989).

The following sections provide more details about the syntax and semantics of MMLOP and give some examples for the application and usability of these concepts.


## 5.2　Syntax

In MMLOP, the symbols '@' and '^' are used to denote 'module' and 'meta' (or 'control') predicates; @lower_case stands for module names (second-order predicates), @UPPER_CASE for predicate variables in module definitions. Parentheses '(' and ')' denote the arguments, whereas braces '{' and '}' delimit the definition of the constructs.

Predicate variables appearing only in the definition of a module but not in the argument list are *internal*; those occurring in the body,

but not in the head of a clause, and in the argument list must be
*imported*; the ones occurring in a clause head and the argument list
can be *exported*.

### Modules

Modules are syntactically characterized in the following way:

- a *module* consists of a *module name*, a list of *arguments*, a *body*,
  which consists of a list of clauses;
- the *module name* (a second-order predicate) starts with '`@`', followed
  by a sequence of `lower_case` letters;
- names for *predicate variables* also start with '`@`', but continue with
  `@UPPER_CASE`;
- predicates to be imported or exported have to appear in the argument list.

A prototypical module has the form

```
@module_name(@IMPORT, @EXPORT) =
   {  <clause_1>
         ...
      <clause_n>
   }
```

### Metaprogramming Constructs

These are denoted by predicates starting with the control sign or '`^`'.


# 5.3   Semantics

A declarative semantics for the above MMLOP program is given by
∀ `@IMPORT` ∃ `@EXPORT` :

    `<clause_1>` ∧ ... ∧ `<clause_n>` → `@module_name(@IMPORT, @EXPORT)`
expressing that for all imported predicates exists an exported predicate
such that the conjunction of the clauses from the definition of the
module logically implies the 'call' of the module with the imported
and exported predicates as arguments.

A *procedural semantics* for the modular part of MMLOP is based on
viewing each module as a separate unit of computation to be executed
according to the underlying evaluation paradigm.  With respect to
LOP and model elimination, a module is a separate LOP program to be
executed by an instance of a LOP evaluation mechanism, e.g. SETHEO

/ PARTHEO. Predicates declared to be imported or exported are subject to *external unification*, denoting that their computation exceeds the limits of modules and requires communication with other affected modules.

The declarative semantics of the *meta-programming* features views meta-predicates (denoted by '^') as second-order predicates, with predicates of the object program as parameters.

A procedural semantics is provided by the concept of *meta-evaluation*; here, the meta-program is evaluated according to the (meta-) calculus[2] used, taking the object program as data for the meta-program and performing the transformations specified by the meta-program on the object program.

## 5.4    Examples

In the following, a few examples are given in order to illustrate the concepts of modules and meta-programming.

**Examples Modules**

The first example (ordered lists) illustrates the use of modules to define abstract data types, whereas the second (parallel theorem prover) shows how to describe the (statical) structure of parallel systems with modules.

**Ordered lists**  of elements are defined by specifying an ordering via an ordering relation ●ORDREL in the module ●order, and, based on this module, another module ●ordered_list with the predicate variable ●ORDERED.

```
●order(●ORDREL) =
 { ●ORDREL(0, s(X)).
   ●ORDREL(s(X), s(Y)) :- ●ORDREL(X,Y).
 }
●ordered_list(●ORDREL, ●ORDERED) =
 { ●ORDERED([X]).
   ●ORDERED([X|[Y|Z]]) :-  ●ORDREL(X, Y), ●ORDERED([Y|Z]).
 }
```

With the additional two clauses (acting as procedure calls)

```
 { ●order(leq).
```

---

[2] or the computational model derived from it, respectively

```
    @ordered_list(leq, ordered).
  }
```

the two modules correspond to the Horn clause program

```
    leq(0,s(X)).
    leq(s(X), s(Y)) :- leq(X,Y).
    ordered([X]).
    ordered([X|[Y|Z]]) :-  leq(X, Y), ordered([Y|Z]).
```

Note that in the modular program @ORDERED is a predicate *variable*, whereas in the Horn clause program `ordered` is a predicate.

**Parallel System Structure**   This example shows the rough structure of a parallel theorem prover, consisting of a number of nodes connected to a host; each node contains a prover, a task store and a communication unit which are interconnected through buses[3]. As an example for a topology, a torus of dimension 4 is given.

```
@partheo(@topology, @SIZE) =
 { @host(0).
   @nodes(@SIZE) :-   @node(1), ..., @node(@SIZE).
 }
@node(I) =
 { @nodestruc(I)  :-
   @prover(I, bus:prover_taskstore),
   @taskstore(I, bus:prover_taskstore,
                 bus:taskstore_communicationunit),
   @communicationunit(I, bus:taskstore_communicationunit).
 }
@topology(@TYPE, @DIM, @SIZE) =
 { @LINKS(I)         :-
       link_1(I), ... , link_@DIM(I).
   @TYPE(torus)    :- @DIM = 4,
       connected(link_1(I), link_3(I+1)),
       connected(link_2(I), link_4(I-4)),
       connected(link_3(I), link_1(I-1)),
       connected(link_4(I), link_2(I+4)).
 }
```

## Examples Meta-evaluation

The meta-evaluation concept is used to specify the execution of modules. The basic principle is to represent the information about the execution behavior of a module in a set of *meta-clauses*, together with

---

[3] note that only the *structure* of the system is specified, not its *behavior*

some (built-in) *meta-predicates* or specific operators (like '|' or '#' for parallel AND).

**Clause Selection**   The selection of a clause from a set of candidates for the extension of the literal under consideration can be directed with the meta-predicate

```
^selclause(LIT, CLAUSE) :- <condition>.
```

A simple application is to give preference to unit clauses:

```
^selclause(LIT, CLAUSE) :- empty_body(CLAUSE).
```

**Literal Selection**   From the current clause, a literal is selected for the next extension step

```
^selliteral(LIT) :- <condition>.
```

Fully instantiated literals are selected by

```
^selliteral(LIT) :- fullinst(LIT).
```

The more interesting case with respect to parallelism is to control sequential or parallel evaluation of the body literals; restricted AND-parallelism, where only literals which do not share variables are treated concurrently, can be described as follows:

```
^restand({LIT_1, ..., LIT_N})
        :- nosharedvars({LIT_1, ..., LIT_N}).
```

**Reductions**   A good application for meta-evaluation is the use of *reductions* to diminish the search space to be traversed.  As an example, pure literals are detected as having no connections, and thus may be discarded:

```
^pure(LIT) :- noconnection(LIT).
```

Meta-evaluation facilities can favorably be used to adapt inference mechanisms to specific applications, enhance them (e.g. via factoring or the use of lemmata), or define new, customized inference machines. Their usability, however, depends crucially on a flexible basic evaluation mechanism, as well as on a good choice and efficient implementation of a number of built-in meta-predicates provided by that mechanism.

### Conclusions

The combination of modularization and meta-evaluation offers an appropriate formalism for the description of complex systems with the means of logic. Their *static* integration in a logic programming system provides features like

- structuring of programs,
- specification of predefined, static problem structures,
- use of abstract data types (code augmentation, "hiding"),
- static configuration control,
- static mapping of modules onto processing elements,
- reductions (preprocessing)

whereas a *dynamic* integration in addition provides evaluation control at run time, e.g.

- modification or choice of the calculus,
- modification of the search rule (clause reordering),
- modification of the computation rule (literal reordering),
- choice of the unification mechanism,
- reductions at run time,
- lemma generation,
- invocation of parallelism,
- load balancing,
- dynamic configuration control,

and many more. Dynamic integration obviously offers a much higher potential, but is also more difficult to implement. For the application programmer it provides very powerful facilities, although it might be difficult to use, or require a different programming style.

# Chapter 6

# Computational Model

The goal of this chapter is to give an idea how MMLOP programs can be evaluated and finally executed on a parallel machine. For this purpose a *computational model* is outlined transforming a MMLOP formula – which may contain some second order predicates – into a set of first order logic formulae, together with some *external* predicates to be unified between elements of this set. The single elements of the set can be executed directly by an appropriate inference mechanism, or be subject to further transformations, e.g. for efficiency reasons, or adapted to a particular inference mechanism.

The execution of a modular logic program which includes meta-programming features is based on a number of evaluation steps. The main idea is to transform the program, which basically represents a second order logic formula, into first order logic. This step is performed via *meta-evaluation*, which in this context means that second order predicates are eliminated by instantiating predicate variables. The result of this step is a set of modules, each described by a set of first order logic programs.

This set of programs is transformed into executable code for the underlying evaluation mechanism (e.g. an abstract machine) by *modularization*. Finally, the actual *execution* is performed, which can be done by an implementation of an abstract machine, or directly by the target system; in the latter case, the result of the previous step must be machine code for that particular system.

## Meta-Evaluation

Meta-evaluation is based on two advanced programming techniques particularly well suited for logic: Meta-interpretation / meta-programming, and partial evaluation. A *meta-interpreter* for a language is an interpreter for that language written in the language itself, and the more general notion of *meta-programming* refers to the treatment of programs as data (Bowen and Kowalski, 1982; Kowalski, 1983; Shapiro, 1986). *Partial evaluation* is a method to compile and object program together with a meta-program (often a meta-interpreter) into a new object program (Jones et al., 1985; Kursawe, 1986; Safra, 1986; Shapiro, 1986; Futamura, 1988; Tanaka, 1988). It maintains the functionality of the original programs, but does not suffer from the overhead usually invoked by meta-programming.

A problem with meta-programming in PROLOG is that the lack of cleanliness in PROLOG's semantics is also apparent if not worse in the meta-programs; especially the use of constructs like **assert** and **retract** for dynamic meta-programming[1] leads to meta-programs with questionable semantics. With the concept of global variables integrated in LOP, the same effects as with **assert** and *retract* can be achieved while maintaining a proper declarative semantics for meta-programs.

Meta-programming, however, can also be used to cure some flaws

---

[1] *dynamic* meta-programming denotes the use of meta-constructs at execution time

of PROLOG; (Sterling and Shapiro, 1986) discusses the definition of an unification algorithm with the occur check; this is a meta-logical problem, since properties of the object program[2] are investigated.

Meta-programs and object programs can be written in one integrated or two separate languages. The use of different languages for meta- and object-programming has the advantage to provide a clear distinction between the two domains; the advantage of an amalgamation is that one single framework covers both domains.

*Meta-evaluation* denotes the usage of meta-programming and partial evaluation techniques in combination, here with the aim of developing an evaluation mechanism for MMLOP.

## Modularization

Modularization designates the decomposition of a program into parts which can be treated independently, with the exception of clearly defined interactions for synchronization and communication purposes. In the case of logic, the modules represent first order predicate logic formulae which share (imported and exported) predicates for the exchange of information.

## Execution

These modules represent largely independent units of computation, to be executed according to the following criteria:

- each module has its own *local execution environment* (which can be a processing element, a process, or a distinguished storage area);
- the *exchange of information* between modules is only possible via imported and exported predicates;
- there is no access to the code of a module from outside.

These criteria are closely related to the concepts of object-oriented programming, and there is some similarity to approaches integrating objects into logic programming (Shapiro and Takeuchi, 1983; Gallaire, 1986; Ohki et al., 1987; Goto et al., 1988). Of particular importance is the fact that there is no inherent need for a central supervisor mechanism which maintains a global overview of the progress of the execution. Thus this model is well suited for architectures without shared memory, e.g. systems based on message passing.

---

[2] here that a variable is not unified with a term containing that variable

# 6.1   A Computational Model for MMLOP

The concepts of meta-evaluation and modularization together with the actual execution serve as basis for a computational model to treat MMLOP programs.

The whole evaluation process is visualized in Figure 6.1, where ovals represent data to be manipulated (the code of the object program), and rectangles stand for transformers performing these manipulations.

The initial point of the evaluation process is the original formula $F$ which may contain second-order constructs, both for meta-programming purposes as well as for structuring the formula into modules. In the first transformation step, *global meta-evaluation*, all meta-programming constructs affecting the whole formula are evaluated as far as possible, resulting in a formula $F'$ which only contains second order constructs dealing with modules and local meta-programming within single modules. In the case that modules are structured hierarchically, this step may have to be applied repeatedly.

The second step is *modularization*, where the transformed formula $F'$ is split up into modules. In the simplest case, only the information given by the module constructs in the program is used; more sophisticated techniques also take additional information into account, e.g. the topology of the target architecture, or additional modules created for alternative spanning sets. The result of modularization is a set of formulae $F_1, \ldots, F_m$, each corresponding to a module, plus predicates to be imported and exported for communication between modules. These formulae still may contain second-order constructs for *local* meta-programming purposes.

In the third step, *local meta-evaluation*, the remaining meta-constructs influencing the evaluation of a single formula $F_i$ are eliminated[3]. The result of this step is a set of first-order formulae, each of which is directly executable by a first-order inference mechanism.

### Example

As an example let us consider a sketch for the evaluation of a large-scale application program, e.g. the one investigated in (Böck, 1989) and originally developed at the Argonne National Laboratory (Argonne, 1988); the application domain is molecular biology, and the problem is to determine the secondary structure of certain ribonu-

---

[3] except for dynamic meta-constructs to be applied at execution time

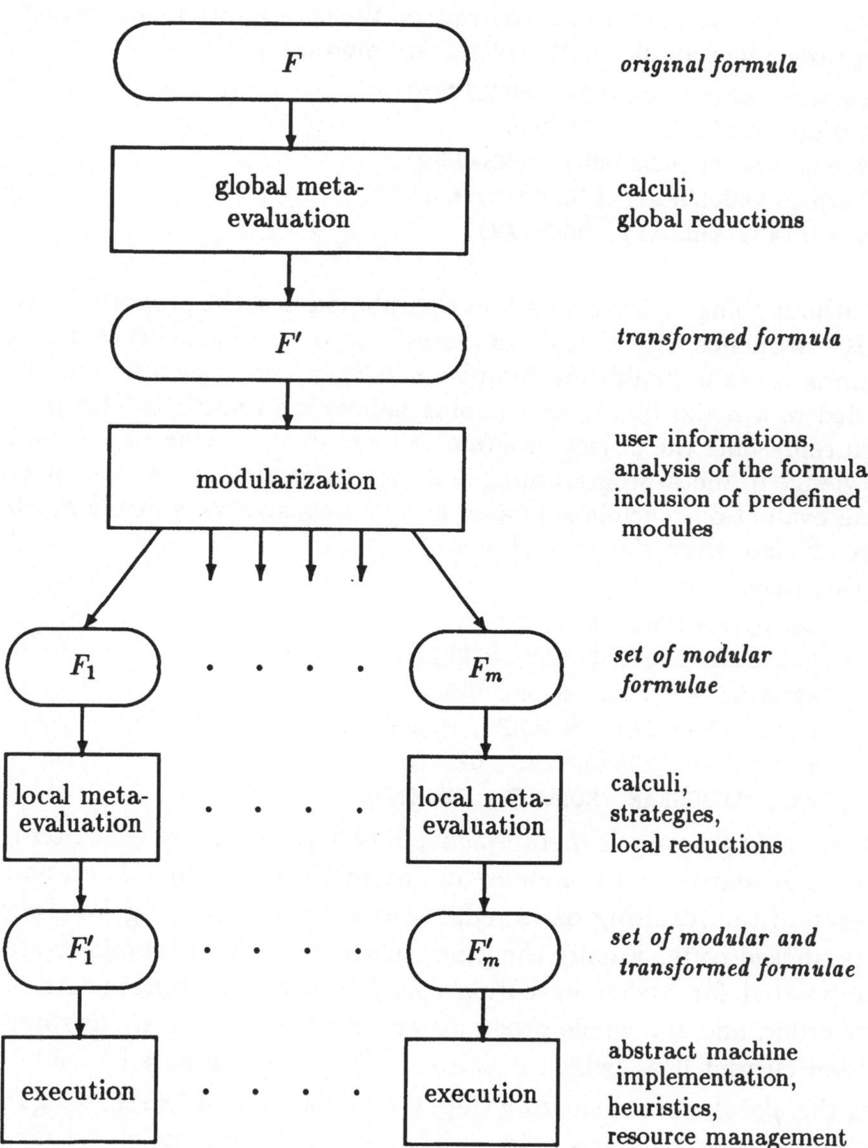



Figure 6.1: A computational model for MMLop

cleid acid (RNA) chains, which means to roughly describe their two-dimensional layout. The whole program consists of three main modules: input, computation, output. Since input and output are not of major interest here, let us concentrate on the computation part, which comprises four modules with several sub-modules each:

```
@determine_structure(@RNA,@HELICES) =
{  @alignment(@RNA, @COLUMNS).
   @complements(@COLUMN1, @COLUMN2).
   @expand(@COLUMN1, @COLUMN2, @HELIX).
   @overlaps(@HELIX1, @HELIX2).
}
```

Without going into the details of the submodules, the program takes a RNA sequence as input, decomposes it into columns, checks if two columns contain significant complements in which case they are expanded to a helix; finally overlapping helices are removed. This program represents the object program to be evaluated. Due to our dual knowledge in meta-programming and molecular biology, we can specify an evaluation mechanism based on meta-constructs which is much more efficient than the general one:

```
^molbiologic(PROGRAM) :-
     ^meta_evac(PROGRAM),
     ^molecularize(PROGRAM, MODULES),
     ^atomize(MODULES, SUBMODULES),
     ^reduce(PROGRAM, MODULES, SUBMODULES),
     ^manipulate(SUBMODULES, GENES),
     ^adapt(PROGRAM, MODULES, SUBMODULES, HOST).
```

The whole program is meta-evacuated (whatever that may be here), then molecularized into modules and atomized into submodules, and on each of the resulting parts reductions are performed, e.g. to eliminate unused code. Finally the genes (clauses) of the submodules are manipulated for higher execution speed, e.g. by rearrangements of their order, and the whole programs and its components are adapted to their current host (which in this case of course is a parallel one).

In the *global meta-evaluation* step, the application of ^molbiologic to **determine_structure** yields a symbiosis of the original object program with the meta-program, which unfortunately is not visible with the naked eye[4]. Since our original program is already nicely structured into modules and submodules, the *modularization* step is almost trivial. In the next phase, *local meta-evaluation*, our knowledge about

---

[4] for more details, you may check out the code with a microscope

internal details of the four modules defined above will be used in order to enhance their evaluation efficiency:

```
^locobio(PROGRAM) :-
    ^filter(MODULE_1).
    ^compromise(MODULE_2).
    ^comprehend(MODULE_3).
    ^oversnap(MODULE_4).
```

Please note that all the above local meta-evaluations are based on *built-in* meta-predicates. In our application it is very important that a particular local meta-evaluation construct is only applied to its specified module; an application of `^filter` to `@complements` might cause obstruction, and `^compromise`ing a `@helix` might even cause `@overlaps` to `^oversnap`. Now that the modules got over local meta-evaluation, they are ready to be executed. This is done with the standard abstract machine based on the molecule elimination mechanism; meta-evaluation during execution is only applied if some helices form a cluster, in which case a use of the `^splinter(HELICES)` facility may be appropriate.

At a first glance, a major restriction of the computational model described above is its assumption that all second-order constructs have to be eliminated before the formulae actually are executed. The transition between the meta-evaluation and the actual execution mechanism, however, is smooth, and at least the local part of meta-evaluation can be integrated into the execution mechanism. Much of the work necessary to evaluate the meta-constructs of a MMLOP formula actually is done at compile time. In principle, meta-constructs can also be applied at run time; such measures may be desirable for the implementation of heuristics, taking into account the current status of the execution, but carry the risk of self-modifying code with all its traps and pitfalls. In cases where extreme efficiency is required, or for special purposes like debugging, load balancing, memory management or performance monitoring, the use of meta-constructs at run-time can be justified.

This computational model for MMLOP is relatively independent of the actual underlying execution mechanism, although some meta-programming constructs can be based on specific features of a particular execution mechanism. As an alternative for the execution on an abstract machine, the construction of spanning sets can be applied; in this case, the executable modules (formulae) are analyzed statically resulting in a number of spanning sets representing potential alternative solutions for each modules. These candidate spanning sets have to be checked for unifiability, and in the case that cycles exist, the necessary copies have to be created.

For efficiency reasons the first order logic programs representing modules can be transformed into Horn clause form by *fanning*, a transformation which creates for each literal of a formula a new clause with exactly that literal as 'entry point' (Letz et al., 1990a). The resulting programs are compiled into abstract machine code (e.g. for SETHEO / PARTHEO), and executed via a simulation, emulation or realization of the abstract machine.

Meta-evaluation provides a technique for the treatment of MMLop programs which is on one hand powerful enough to express the exploitation and control of parallelism in an adequate way, and on the other hand efficient enough to deal with implementations of non-trivial problems on parallel computers. Whereas higher order logic may not always be appropriate due to its complexity (Joseph, 1988), modularity or something equivalent is a very essential way to structure parallel programs, both with respect to the initial structure of the problem and the architecture of the target system.

# Chapter 7

# Architecture

In this chapter, important architectural aspects for the design of a parallel inference machine are outlined. The basis for this discussion is provided by mappings of the spanning set approach as execution mechanism for logic programs onto different types of parallel computer architectures. This approach is particularly suited for parallel evaluation because it identifies independent computational units at compile time, does not rely on shared data structures, separates the treatment of logical relations (predicates) and structure relations (terms), and can be combined with other categories of parallelism (reductions, competition, term parallelism). Its maintenance of a clear declarative semantics together with good prospects for efficient execution make it a suitable candidate as underlying execution mechanism for a language like MMLOP. In addition, possibilities for parallel implementations of unification, especially aiming at dedicated coprocessors, are investigated.

The language MMLOP and its underlying computational model provide powerful mechanisms for the description and evaluation of parallel problems. The logical basis offers a high expressiveness together with a well-founded theory, and a declarative as well as a procedural semantics can be defined without the use of non-logical features as in PROLOG. A crucial factor for practical usability is also the efficiency of the evaluation mechanism which still is a problem for logic-based systems. The goal of this chapter on architecture is to outline a basic execution mechanism for logic programs which has a clean semantical foundation together with the potential for good performance.

The basic execution mechanism proposed here is a derivative of the spanning set approach. This approach is based on a static analysis of the formula to be evaluated, as described in Chapter 2 and Section 4.8. The outcome of the analysis are spanning sets, specifying alternative solutions for the formula (OR-parallelism); the analysis can also be extended towards a detection of routes, which represent sequences of connections required to compute a solution for one spanning set. On this basis, units of computation are identified which can be executed totally independently (the different spanning sets), or as independent as the problem formulation allows (routes, restricted by the occurrence of shared variables). Such an analysis at compile time allows a much better usage of parallelism than most parallelizations of PROLOG, which are based on an execution model where parallelism occurs dynamically. Dynamicity, of course, also appears in the spanning set approach, but is restricted to cases where it is explicitly specified in the user program.

The other important feature of the spanning set approach is the extraction of structural information via a concentration of the terms to be unified into one huge term pair per spanning set. As a consequence, the exploitation of fine grain parallelism on the term level is much more promising than in resolution-based approaches where only the terms of a pair of literals are unified in one step. Due to the importance of unification, especially for the spanning set approach, efficient mappings of unification mechanisms onto different types of parallel architectures are investigated as well; some of these mappings are based on specialized unification coprocessors attached to conventional processors.

# 7.1  Overview Architecture

In the sequel, the relations between the investigation of different categories of parallelism in logic, the parallel logic language MMLOP together with an underlying execution mechanism based on spanning sets, and parallel computer architectures as implementation vehicles for such a mechanism are discussed.

The following section deals with the functional structure of an execution mechanism based on the spanning set approach; then the co-operation of a number of these mechanisms is sketched, followed by mappings onto bus-based, array- and tree-structured, and massively parallel architectures. The last section of this chapter resumes a unification mechanism presented in (Bibel et al., 1987); it uses a dag representation of terms to be unified, and is suited for implementation on small grain parallel devices as well as on processing elements like Transputers.


# 7.2  Spanning Setters

A *setter* is a long-haired dog that can be trained to show hunters where birds and animals are; a *spanning setter* is one trained to detect spanning sets of connections and cycles in a logical formula[1]. The action phase of a spanning setter typically adheres to the following pattern:

- ferret out important information in the formula (e.g. connections);
- gather connections into candidates for spanning sets;
- validate the eligibility of the candidates.

The actual appearance of the components making up such a pattern of course depends on the background of the spanning setter in service; essential factors are for example the hunting philosophy (the calculus applied) and the validation method for the eligibility test (where full unification should be aspired).

In former times, spanning setters had to pursue their affairs in the thicket of full first order logic without much help; after some time, they learned to make use of connection graphs for better guidance,

---

[1] no real animals have been used for the experiments; all results could be achieved by computer-animated evaluations

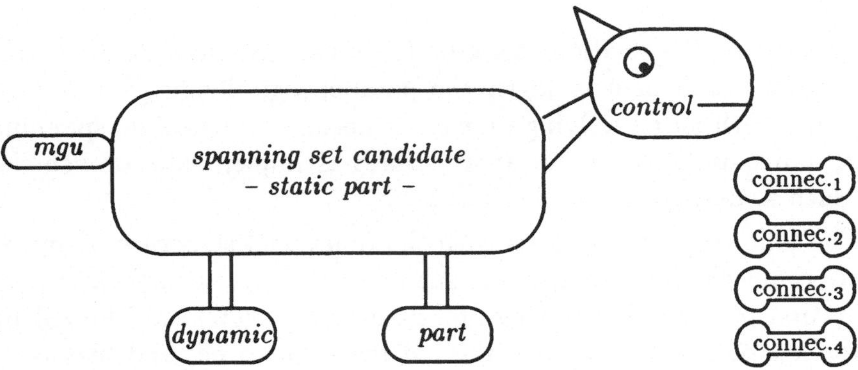


Figure 7.1: Anatomy of a spanning setter

and after some domestication they are well-adjusted to routes in the meantime.

The physical aspects of a spanning setter reflect its adaptation to its main task of hunting down logical formula: as soon as an initial connection is found out, the control unit initiates the appropriate actions to construct the spanning set candidate; after the static part is available, the dynamic part is used to trace the *mgu* which finally decides upon the usability of the spanning set candidate. Figure 7.1 gives an outline of a spanning setter's anatomy.

The hunt for spanning sets is a good occasion to make use of parallelism; a general (maybe a bit too generous) strategy is to put one spanning setter on every spanning set candidate. The hunting strategy of a team of spanning setters is illustrated in Figure 7.2.

The requirements on an architecture for a parallel machine based on the spanning set approach are discussed in the following. The basis is the specification given in Section 4.8; Figure 7.3 resumes the functionalities required for the spanning set approach:

**Determination of connections:** The connections occurring in the formula to be evaluated have to be found; the basic operation is to parse the textual representation of the formula for pairs of complementary literals. It is a relatively simple task, typically done in a very early stage of the evaluation, storing the connections found in a list or array for later usage. Due to its simplicity and the necessity of global information (the whole formula), it is in general question-

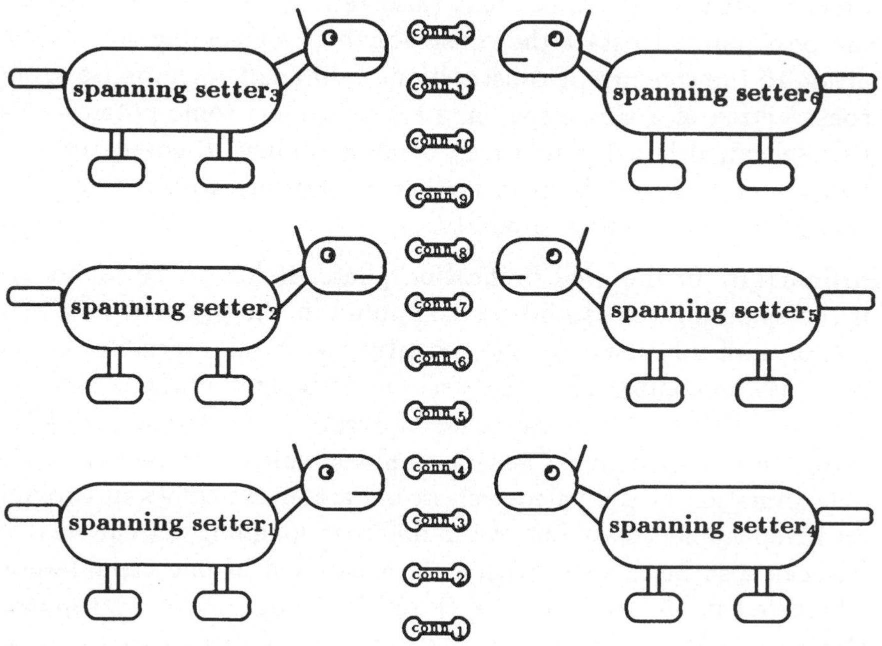



Figure 7.2: A team of spanning setters

able to parallelize this task; only in the case of modularity, where a formula consists of or can be partitioned into several modules, these modules obviously may be treated concurrently. In order to avoid redundancies in the next stage, the connections found can be classified according to their suitability to serve as initial connections for the generation of spanning sets.

**Generation of spanning set candidates:** Spanning set candidates are constructed by incrementally adding appropriate connections; for a candidate set to be spanning, every path in the matrix must have a connection in the set. In addition, potential cycles in the matrix have to be found for a later extension of the set. So the basic operations are the construction of a set with certain properties, which can also be viewed as finding a particular sub-graph in a graph represented by the connections of the formula. With respect to parallelization, an extreme approach would be to start from each connection of the formula and try to find the corresponding set; this, of course, can be extremely redundant since a spanning set candidate typically contains more than one connection. Different spanning sets for a formula exist if there are literals where more than one connec-

tions emerge; these connections (also termed OR-connections) are
the best points to start the construction of a spanning set. In addition to the concurrent construction of different spanning sets, the
construction of a single spanning set also offers some potential for
parallelism, although a relatively small grain one. It emanates from
two sources: the addition of further connections to a set, and the
check for the 'spanning' property.

**Unification:** In the final unification phase, it has to be found out
if the spanning set candidates computed in the previous step represent valid solutions, i.e. the substitutions for the variables in one
set must be compatible. The basic underlying operations here are
the comparison of the structure of terms to be unified, and the
computation of the most general unifiers, which corresponds to the
determination of equivalence classes. Parallelism comes in through
an independent treatment of the different spanning set candidates,
but can also be invoked within the treatment of one candidate as
advocated in (Bibel et al., 1987) and implemented on Transputers
(Hager and Moser, 1989a); in addition, function evaluation and occur check can be done in parallel, and terms may be traversed in a
bottom-up way.

Related approaches, but on a more abstract level, are presented
in (Ibañez, 1989; Balcke, 1987); the matrix method (Prawitz, 1970),
which shares some important features with the connection method,
is the basis of a compiled approach to the execution of Horn clause
programs (Wang, 1989).

Let's examine the functional parts of Figure 7.3 in more detail. The
*determination of connections* is done in an early phase of the evaluation of a formula (e.g. during input, compilation[2], or preprocessing).
Since it is not very interesting from a parallel processing viewpoint,
we will not go into details and just assume that as an outcome a list
of connections will be available, where OR-connections are especially
marked. The actual computation of spanning sets is done concurrently
by a number of identical functional units called *spanning setters*. For
the sake of simplicity let us assume that there is one spanning setter
for each OR-connection of the formula, serving as crystallization point

---

[2] if an approach is used which compiles the formula into the code of an abstract
machine

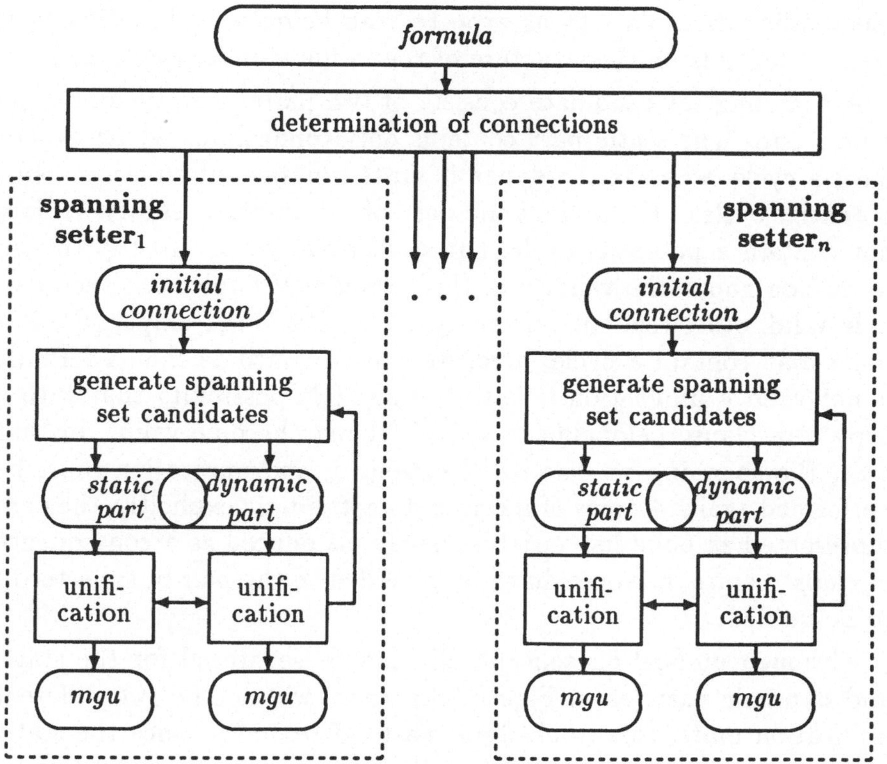


Figure 7.3: Functional structure of spanning setters

for a spanning set candidate[3].

Starting from an OR-connection, a spanning set candidate is determined by incrementally adding connections to the set until the set contains a connection from each path of the formula. For the construction of such a candidate set, it is advisable to follow some strategy in order to avoid the addition of connections which do not contribute anything. A strategy can be derived from a logical calculus (e.g. connection method, model elimination, resolution) by postponing unification, or it can be especially designed for a particular approach (as in (Balcke, 1987; Ibañez, 1989)). The latter has the advantage that a number of important features can be determined by a statical analysis of the formula, thus facilitating the further evaluation, but suffers from the overhead induced by the analysis. Despite this overhead, we favor a specialized solution since the performance gain – especially for

---

[3] otherwise, some spanning sets have to be treated one after the other, or in a multitasking way

'demanding' formula – to be expected can be quite high, although it greatly depends on the structure of the formula under evaluation.

A spanning set candidate consists of two parts, a *static* and a *dynamic* one. The static part contains only connections which cannot form a cycle, whereas the dynamic parts contains information about potential cycles. If the dynamic part of a formula is empty, it does not contain a potential cycle, and unification yields directly the information about the validity of this formula: if unification succeeds, it is valid, otherwise not. If the dynamic part is not empty, the formula may contain a cycle; here, we cannot conclude that a formula is not valid if unification fails since there is a possibility that with a copy of a clause belonging to a cycle it may become valid. In that case, the spanning set has to be extended, and unification must be performed again. Copies of a clause do not appear explicitly, they are represented as pointers; variables newly introduced as a consequence of clause copies, however, have to be added to the pair of superterms to be unified.

Although unification conceptually can be separated for the static and dynamic part (as in Figure 7.3), there will not be two different unification units; this would be a waste of resources since the static one would only be invoked once, and the dynamic one would only be used after the static one delivered **false** as result.

Of equal importance as a strategy for the construction of spanning set candidates is an appropriate choice of the data structures used to represent the formula, here especially the term structures of the literals. Since unification is postponed until the static part of a spanning set candidate has been located, a separate representation for the single literals and their terms is not very well suited. It is better to combine all the terms comprised in the connections (which are pairs of complementary literals) of a spanning set candidate into one huge pair of superterms to be unified in the unification phase (as advocated in (Bibel et al., 1987).

The pair of superterms can be represented as a *dag* (*d*irected *a*cyclic *g*raph), with the advantage that multiple occurrences of symbols (especially variables) are represented only once[4]. With respect to the

---

[4] this is of importance especially for the treatment of shared variables, facilitating the maintenance of consistent bindings and avoiding additional consistency checks as in Delta PROLOG (Moniz Pereira et al., 1988; Cunha et al., 1989), for

underlying architecture, the representation and treatment of terms to be unified is even more important than the strategy for the construction of spanning sets since different representations can require different architectures.

The architecture of the unification unit proposed here is based on a dag representation for the terms, and the concept of forwarding the connections until terminal symbols are reached, followed by the computation of equivalence classes to determine the variable substitutions; in addition, an occur check is performed to avoid cyclic substitutions (see (Hager and Moser, 1989a) for details). In the same paper it has been shown that a network of Transputers can be used efficiently for such a unification mechanism; as an alternative, a special purpose unit based on associative memory (Robinson, 1986; Georgescu, 1986) looks promising. Another alternative which is especially suited for 'sloppy' or 'fuzzy' unification[5] lies in the use of a neural network (Kurfeß and Reich, 1989); in a different context, the use of neural networks for unification is discussed in (Stolcke, 1988; Hölldobler, 1990b; Kurfeß, 1991).

# 7.3   Mappings to Parallel Architectures

One of the reasons to use a parallel specification language like UNITY is to provide a good foundation for the systematic development of programs for a variety of architectures and applications. The main theme of this section is to identify requirements imposed onto an implementation of the spanning setter concept on different types of parallel systems, together with additional categories of parallelism incorporating coarse (e.g. modules, competition, reductions) or fine grain (e.g. unification) computations. These mappings are investigated according to a number of important issues on the node and systems architecture level, as well as operational characteristics of the system as a whole. Table 7.1 gives an overview of important features for parallel architectures in general, together with some keywords for typical or possible realizations.

A restriction on the investigation done here is that only parallel sys-

example
[5] unification where a certain degree of inconsistency is allowed; this is especially important for real world knowledge representation and processing

<table>
<tr><td colspan="1" align="center">PARALLEL ARCHITECTURES</td></tr>
</table>

| PARALLEL ARCHITECTURES |
| --- |
| *System Architecture* |
| **interconnection topology:** fixed / flexible; point-to-point connections, switching networks, bus<br>**homogeneity:** homogeneous / heterogeneous<br>**expandability:** integration of additional nodes<br>**memory structure:** separate / shared / distributed, global address space |
| *Node Architecture* |
| **structure:** typically processor + memory, sometimes separate communication units and / or special purpose coprocessors<br>**functionality:** general purpose / specialized, adaptable (microprogrammable) |
| *Operational Characteristics* |
| **control:** central / distributed, one level / hierarchical, master-slave / equal rights<br>**topology:** partially / (virtually) fully interconnected, fixed (compile time) / flexible (run time)<br>**communication:** message passing, shared data, global address space<br>**synchronization:** messages, remote procedure call, bus access, shared memory, special hardware<br>**process switching:** important for multitasking (more processes than processors)<br>**granularity:** coarse / medium / fine<br>**load distribution:** fixed, static (at compile time), dynamic (at run time); centralized, distributed<br>**memory management:** virtual memory, garbage collection<br>**debugging:** post-mortem, single-step mode, dynamic, deterministic re-run (instant replay), special hardware |

Table 7.1: Features of parallel architectures

tems are discussed, but not *distributed* ones, like networks of workstations, for example. Provided that a set of basic language or operating system primitives for distributed computation (e.g. remote procedure calls, or LINDA (Gelernter, 1988)) is available, there are no fundamental obstacles preventing a realization of the spanning setter concept on a distributed system.

## Bus

In the sequel, the essential features of an implementation of the spanning setter concept on a bus-based architecture are outlined. Figure 7.4 shows an overview of a bus-based system, with four alternative units connected to the bus. The purpose of the figure is not to show a realistic configuration, but just to give an impression of the components a bus-based system for an implementation of the spanning setter concept might be composed of.

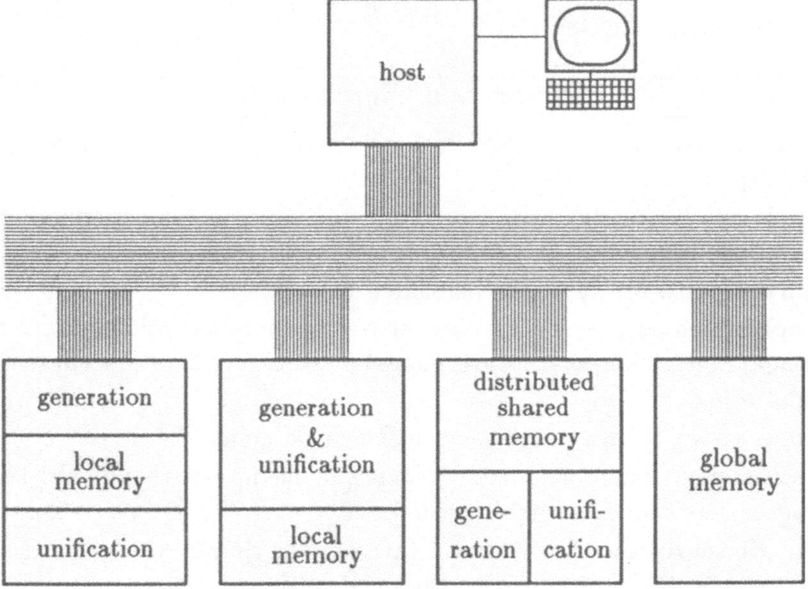


Figure 7.4: Bus-connected spanning setters

A bus consists of a physical connection of all units in the system to a number of common lines for the transmission of data and control signals. This facilitates an easy exchange of information, e.g. by sending messages to certain nodes, establishing mailboxes, or by providing a global address space.

A major problem of bus-based system is the limitation of the bus access to only one node at a time; this enforces the establishment of an access protocol and restricts the available bus bandwidth. As a

<table>
<tr><td colspan="1" align="center">Bus System</td></tr>
</table>

| Bus System |
| --- |
| *System Architecture* |
| **interconnection topology:** bus; spatially fully connected, but the access is temporally restricted to one node; requires access protocol (token, time slices, collision detection)<br>**homogeneity:** heterogeneous nodes possible, only bus specifications have to be met<br>**expandability:** easy, but limited by physical and access limitations<br>**memory structure:** local, distributed, shared possible, global address space; simultaneous write must be prevented |
| *Node Architecture* |
| **structure:** apart from the bus interface no specific requirements<br>**functionality:** flexible |
| *Operational Characteristics* |
| **control:** central (bus master) or distributed (managed via the bus protocol)<br>**topology:** (virtually) fully interconnected, flexible (at run time), but only one connection at a time<br>**communication:** typically via shared memory<br>**synchronization:** bus access and shared memory as basic mechanisms; semaphores, monitors, (synchronous) messages, remote procedure call, and others on top<br>**process switching:** dependent on the single nodes<br>**granularity:** medium to large, dependent on the bus (bandwidth) and the nodes (process switching time)<br>**load distribution:** fixed, static (at compile time), dynamic (at run time) possible; limitations by the bandwidth<br>**memory management:** virtual memory, background memory, local and global garbage collection<br>**debugging:** post-mortem, single-step mode, dynamic via bus monitor |

Table 7.2: Bus-based system

consequence, only a relatively small number of nodes can be connected to a single bus, typically up to twenty or thirty. Approaches to solve this dilemma are to increase the width of the bus, the use of multiple buses, or clustering large systems into interconnected smaller ones.

Buses are technologically well established from conventional architectures, and the necessary concepts for multiple processing units can be implemented with moderate effort. Thus their use becomes more and more popular in mid-size systems, especially in combination with the UNIX operating system. Table 7.2 resumes important characteristics of bus-based systems.

The spanning setter concept can be implemented on a bus-based system in a quite straightforward way. The determination of connections is done on the host, and the connections are distributed to the single nodes, serving as initial connections for the generation of a spanning set candidate (cp. Figure 7.3). For the generation, information about the structure of the whole formula is required. This information can either be stored in a global memory and accessed by need from every node, or copied into the local memories of each node. Since the amount of information which has to be transferred is relatively small, the latter alternative is preferable. The actual computation and validation of the different spanning sets are independent for each spanning set, and the bus load can be held quite low. For tasks with large or medium grain sizes, a bus-based architecture is appropriate as implementation vehicle for the spanning setter concept as well as for many other evaluation mechanisms for logic programs. For tasks with finer grain sizes, most existing systems are not very well suited; one of the reasons for that is the relatively high task switching time on conventional processors in comparison with the Transputer, for example.


## Transputer Systems

The following discussion centers around the use of Transputers (INMOS, 1986; INMOS, 1988a) as components of a parallel system to implement the spanning setter concept. In terms of parallelism, Transputers provide building blocks to easily compose multi-processor systems connected in a point-to-point fashion through links, which establish channels to transfer information in the form of messages. Figure 7.5 shows a possible structure of a single node including a dedicated unification coprocessor; design alternatives for that are discussed in Section 7.4.

A rough outline of a multi-processor system based on Transputers is

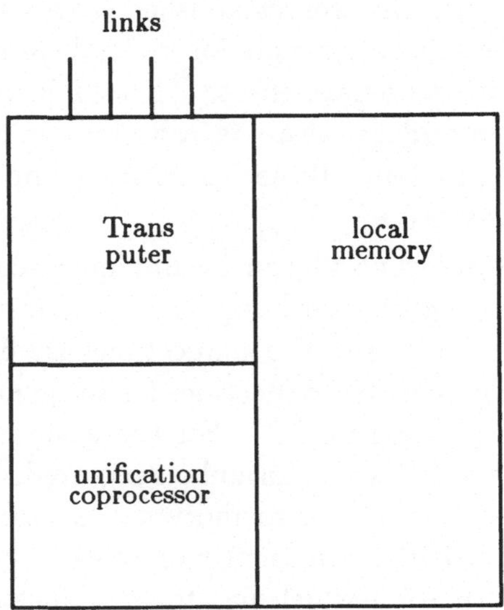



Figure 7.5: Transputer-based node

given in Figure 7.6; the dots between the links are to indicate that the actual topology is not necessarily a grid, but can be changed either by the wiring or by link switches.

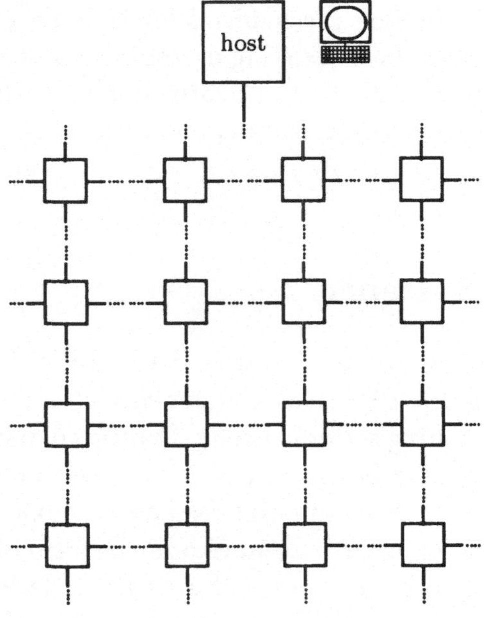



Figure 7.6: Transputer system

An instance of a particular topology is presented in Figure 7.7; it is an extension of the master-slave concept used for parallel unification on Transputers as described in (Hager and Moser, 1989a).

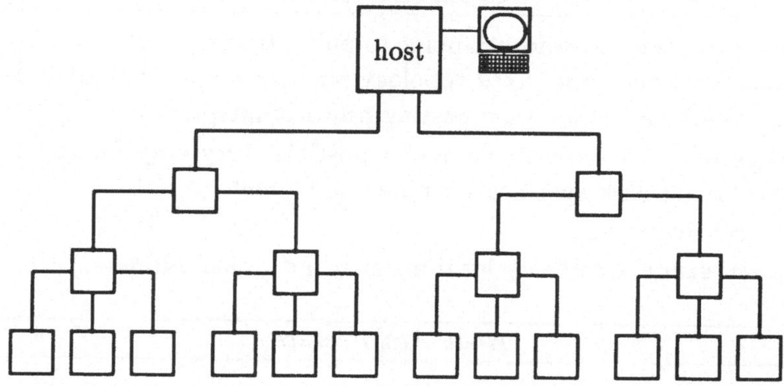


Figure 7.7: Transputer tree

The essential characteristics of parallel systems based on Transputers are shown in Table 7.3. In such a system, a node can be directly connected to $n$ neighbor nodes, where $n$ stands for the number of links a node has; for the current Transputer generation, $n$ is 4. Consequently, a physically fully interconnected network is not feasible for larger systems; link switches are used to change the topology without rewiring the system, and software mechanisms like routing accomplish transfer of information between nodes which are not directly connected.

In an implementation of the spanning setter concept on Transputers, the determination of the connections is also done on the host, and the connections distributed to the single nodes for the generation of spanning set candidates. Since the exchange of information between the single nodes is relatively slow due to the serial links, information needed by all nodes should be kept in the local memories, if possible; with an appropriate representation, this is no problem here.

Although it is not ideal for that purpose, a Transputer-based system can even be used for a small grain task as parallel unification (Hager and Moser, 1989a). Their fast task switching mechanism also allows reasonable usage of multitasking within the single node, which facilitates the mapping of problems incorporating more tasks than there are processing elements in the system. Together with the relative small costs per processing element, Transputers present a very good basis for an implementation of logic evaluation mechanisms with independent computational tasks, such as the spanning setter concept, and

| TRANSPUTER SYSTEM |
|---|
| *System Architecture* |
| **interconnection topology:** point-to-point (links); restricted to four connections per node; fixed topology, unless switches for the links are used; synchronous message passing protocol integrated <br> **homogeneity:** heterogeneous nodes possible, very easy for the Transputer family; link specifications have to be met <br> **expandability:** easy <br> **memory structure:** only local memory, no global address space; |
| *Node Architecture* |
| **structure:** apart from the link interface no specific requirements; very easy to build <br> **functionality:** good with relatively little effort |
| *Operational Characteristics* |
| **control:** hierarchical (master – slaves) or distributed (managed via messages) <br> **topology:** (virtual) full interconnection possible via link switches or message routing, flexible (at run time), simultaneous transmission on all links in the system <br> **communication:** message passing <br> **synchronization:** synchronous messages <br> **process switching:** very fast <br> **granularity:** mainly medium to large, but also suited for some small grain tasks <br> **load distribution:** fixed, static (at compile time), dynamic (at run time) possible, but difficult; limitations by the bandwidth of the links to be transferred <br> **memory management:** no support for virtual memory; background memory on special nodes (servers), global garbage collection only via links <br> **debugging:** post-mortem, single-step mode; dynamically quite difficult: no central instance with overall system information, access to the single nodes only via links; instant replay monitor (Noske, 1989); sometimes special hardware (service bus) |

Table 7.3: Transputer-based system

have already been used successfully for the implementation of a similar inference mechanism, PARTHEO (Schumann and Letz, 1990; Ertel et al., 1989).

## Massively Parallel Systems

The notion of massively parallel systems comprises architectures composed of a huge number of relatively simple processing elements, each consisting of a small processor, some local memory, plus some communication facilities; representatives of these systems are the Connection Machine (Hillis, 1985; Thinking Machines, 1987), the Massively Parallel Processor (MPP) (Potter, 1985) the Distributed Array Processor (DAP) (Hwang and Briggs, 1984), and a number of others are under investigation or construction.

Due to the synchronous (SIMD) operation of these systems, a different programming style must be applied: in each step, one atomic operation is executed in each processing element on the corresponding data item; depending on simple conditions, the operation can also be suppressed, but it is not possible to execute different operations at the same time. This results in the use of large regular data structures (often arrays) and extensions of programming languages to treat such structures synchronously (Hillis and Steele, 1986). For problems which can be mapped onto such regular structures, the performance of massively parallel systems can be very high (Frenkel, 1986). An impression of a massively parallel system is given in Figure 7.8.

The implementation of spanning setters as well as other logic evaluation mechanisms on massively parallel systems requires major changes in the coding with respect to conventional systems, and in many cases in the whole evaluation concept.

The mapping of programs specified in UNITY is quite straightforward for those parts making use of the assignment statement ||: all the components of an assignment are executed independently and simultaneously by evaluating their righ sides and assigning the resulting values to the corresponding variables on the left sides. A problem might be, however, that some complex constructs allowed on the right side of UNITY assignments cannot be realized easily on the simple processing elements of massively parallel systems. In some cases it can be difficult to find an appropriate representation of the data struc-

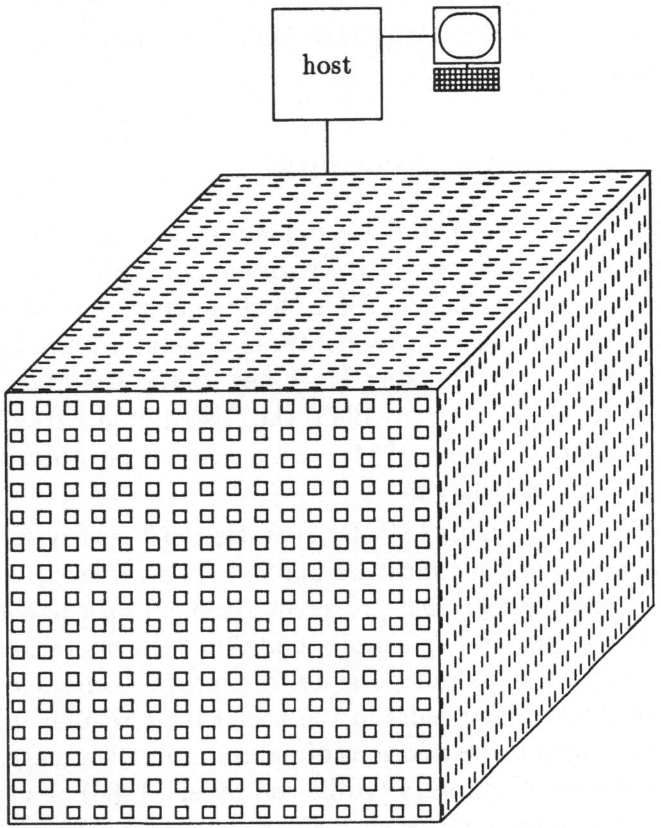


Figure 7.8: Massively parallel system

tures, especially since so many items of the same basic type must be arranged in a regular way.

In principle, the spanning setter concept can be executed synchronously: assuming that (candidates for) spanning sets are represented as graphs, their constructions corresponds to a computation of the transitive closure of that graph which can be computed synchronously by adding all connected nodes to the current set of nodes. In a similar way, unification can be performed by forwarding information about subterms to be unified.

The evaluation of logic programs according to the common logical calculi does not result in a regular execution pattern which can be easily transferred into synchronous operations. The problem which arises for the usage of massively parallel systems in this area is that it might be difficult to make use of some thousands of processing ele-

<table>
<tr><td colspan="1" align="center">MASSIVELY PARALLEL SYSTEM</td></tr>
<tr><td align="center">System Architecture</td></tr>
<tr><td>

**interconnection topology:** point-to-point (links); restricted to relatively few connections per node unless switches for the links are used
**homogeneity:** heterogeneous nodes difficult
**expandability:** difficult (system-specific)
**memory structure:** local memory, global address space (processing elements);

</td></tr>
<tr><td align="center">Node Architecture</td></tr>
<tr><td>

**structure:** specialized processing elements, local memory, sometimes communication unit
**functionality:** simple, operations on bits

</td></tr>
<tr><td align="center">Operational Characteristics</td></tr>
<tr><td>

**control:** central (synchronous / SIMD)
**topology:** (virtual) full interconnection possible via link switches or message routing, flexible (at run time), simultaneous transmission on all links in the system,
**communication:** message passing
**synchronization:** synchronous mode of operation
**process switching:** depending on the processing element
**granularity:** small
**load distribution:** fixed, synchronous mode of operation
**memory management:** background memory on the host or through special nodes (servers),
**debugging:** post-mortem, single-step mode; through the host (central control instance)

</td></tr>
</table>

Table 7.4: Massively parallel system

ments, and at the same time avoid uncontrolled combinatorial explosion. Some calculi, however, which are not very interesting for human usage or for implementation on conventional systems might be suitable for massively parallel systems.

# 7.4    Unification

Unification being *the* basic operation during the evaluation of logic programs, it typically consumes a lot of the required computation time. Substantial reductions can be accomplished either through specialized devices, or through parallelism in the unification process, or a combination of both. A parallel unification mechanism has been outlined in (Bibel et al., 1987), and subsequently implemented on Transputers (Hager and Moser, 1989a); the system topology presented previously in Figure 7.7 is an outcome of that work.

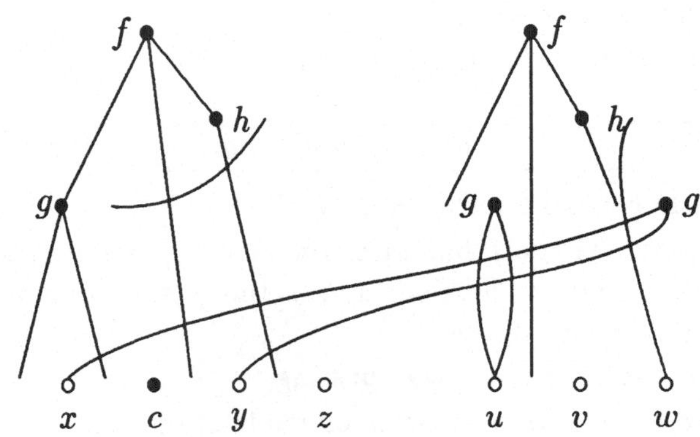



Figure 7.9: Dag representation of terms

Let's consider the unification of the two terms $f(g(x, c), y, h(g(x, c), z)$ and $f(g(u, u), v, h(w, g(x, y))$ as an example; a dag representation is given in Figure 7.9[6].

The initial situation for the unification process is shown in Figure 7.10; the dashed arc denotes that the substructures – in this situation the two entire terms – connected by it have to be unified.

In Figure 7.11, the arc has been successfully forwarded to the successors of the two root nodes, and three new arcs have been created. One of the three arcs already contains terminal nodes (the two variables $y$ and $v$), and the (partial) substitutions for these variables can be computed. In the case that an arc contains a terminal node, it is called *terminated* arc.

---

[6] note that the sub-term $g(x, c)$ appears twice in the first term, but is only represented once in the dag; such a representation, where shared substructures appear only once, is also referred to as *minimal dag*

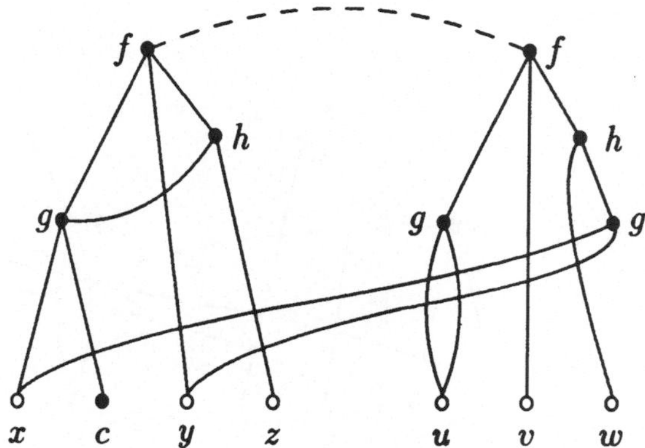

Figure 7.10: Root connection

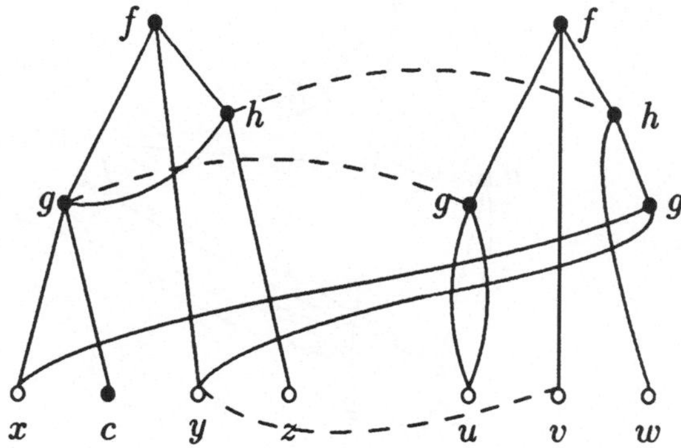

Figure 7.11: First forward

The end of the forwarding phase has already been reached after two steps: each arc touches at least one terminal node (see Figure 7.12).

After the forwarding, the computation of the substitutions must be done, which can be depicted as 'melting' the nodes connected by arcs (see Figure 7.13). The melting of the nodes obeys the rules known form unification: variables assume the values from constants and functions, a melting of uninstantiated variables remains uninstantiated, while constants and functions can only melt with the same type and if they have the same symbol.

As soon as the forwarding phase has come to an end, the occur check can be performed, inspecting the term structure for potential cycles.

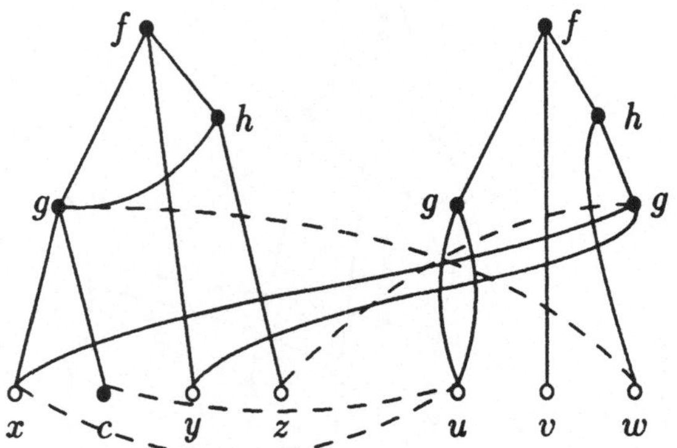

Figure 7.12: Second forward

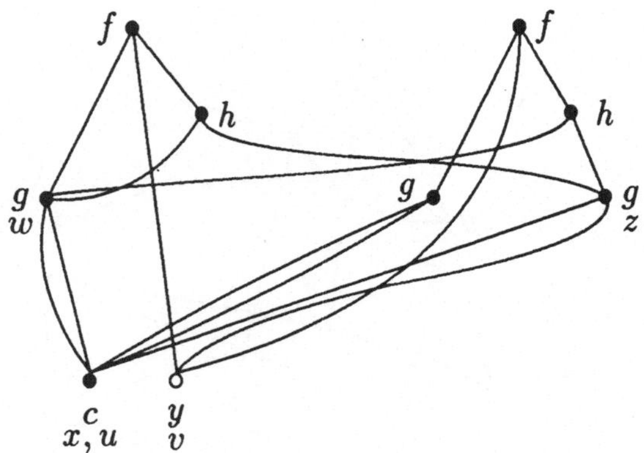

Figure 7.13: Melted dag

Some alternatives to be used as candidates for specialized unification coprocessors in the respective functional units of Figures 7.5 and 7.4 are

- microprogrammable devices,
- massively parallel systems (Connection Machine, MPP, DAP)
- systolic arrays (WARP),
- associative memories, and
- neural networks.

Their suitability for the execution of unification according to the forwarding concept with a da representation of terms described above is briefly discussed in the following paragraphs.

### Microprogrammable Devices

Systems with microprogrammable processing elements can be used to efficiently implement unification algorithms, largely independent of the particular algorithm and data representation chosen. Of special interest could be a combination of microprogramming and associative processing, as advocated (for general purpose processors) in (Albert and Bode, 1983; Bode, 1980).

### Massively Parallel Systems

Parallel computer systems with a huge number of relatively small processing elements are suitable for small grain operations on many data items of one and the same type (Hillis and Steele, 1986); for term unification, these large machines of course are much oversized. The operational principle, however, can be applied to a dedicated unification coprocessor, consisting of maybe a few hundred simple processing elements. In this case, the size of the terms still has to be sufficiently large, and an approach as described in this chapter with a separation of the unification phase and an integration of all term pairs into one huge pair of super-terms represented as dags is more appropriate than the step-by-step unification of small term pairs typically used in conventional inference mechanisms. Program SYNC-UNIF (Figure 7.14) describes the synchronous treatment of a dag structure to be unified; the terms are represented as an array, and in each step all the nodes whose status are **active** forward the unification task to their successors by informing them about their respective unification partners. In the case that either a node or its partner represents a terminal node, the two are melted (provided that this is possible).

### Systolic Arrays

Instead of storing the nodes of the dag in processing elements and manipulating pointers to achieve forwarding, a pair of terms can be unified in a systolic way by feeding the two terms into the array step by step such that corresponding subterms are treated by the same element of the array; for a general description of the systolic array concept see (Quinton, 1987), for example. A problem here is that a systolic array requires inputs of the same width[7] in each step, whereas the breadth of a term[8] may vary from level to level. This requires a proper alignment of the two terms before the actual processing in the

---

[7] the same number of data items
[8] the number of nodes on a level

**Program SYNC-UNIF**
**parameters**

| *name* | *type* | *initially* | *description* |
|---|---|---|---|
| *dag* | array of nodes $n_i$ | | pair of terms to be unified |
| $n_i$ | (smb, type, succs, ptn) | | starting nodes |
| $succ_j(n_i)$ | node | | $j$-th successor of node $n_i$ |

**local variables**

| | | | |
|---|---|---|---|
| $stat(n_i)\{$**passive, active**$\}$ | **passive** except for one | status of a node | |
| $ptn(n_i)$ pointer to a node | **nil** except for one | unification partner of a node | |

**computation**

$\| \ \langle \quad i,j : \ stat(n_i) = \text{active}, n_{ij} = succ_j(n_i) :: \qquad$ for all active nodes and their successors

$\quad \| \ ptn(succ_j(n_i)) \quad := \ succ_j(ptn(n_i)) \qquad$ forward unification to
$\qquad\qquad \text{if } type(n_i) = \text{nonterminal} \qquad$ the successor nodes
$\qquad\qquad \wedge \ type(ptn(n_i)) = \text{nonterminal}$

$\quad \| \ stat(succ_j(n_i)) \quad := \ \text{active} \qquad$ mark successor nodes
$\qquad\qquad \text{if } type(n_i) = \text{nonterminal} \qquad$ as active
$\qquad\qquad \wedge \ type(ptn(n_i)) = \text{nonterminal}$

$\quad \| \ melt(n_i, ptn(n_i)) \qquad \text{if } type(n_i) = \text{terminal} \qquad$ melt nodes
$\qquad\qquad \vee \ type(ptn(n_i)) = \text{terminal} \qquad$ if one is terminal

$\quad \rangle$

**end SYNC-UNIF**

Figure 7.14: Program SYNC-UNIF

array, and it is questionable if complete unification could not be done with little additional effort.

**Associative Memories**

The basic idea for unification with associative memories is to store one term as pattern in the memory, present the other as input pattern to the memory, and find out if they are compatible (Kohonen, 1987; Palm, 1980; Jackoway, 1984). This task is relatively easy if the operation to be performed is not full unification, but restricted to matching. Nevertheless associative memories are of interest for the evaluation of logic programs because they also can be used for tasks other than unification which involve a comparison of structures, e.g. selection of candidate clauses. (Robinson, 1986; Georgescu, 1986).

**Neural Networks**

One well-suited application for neural networks are pattern comparisons; with a suitable representation, a pair of terms to be unified can be presented to a neural network as a pair of patterns to be compared (Kurfeß, 1991; Hölldobler, 1990b; Kurfeß and Reich, 1989; Stolcke, 1988). Similar as in associative memories, full unification is difficult to achieve; in addition, problems can arise with the representation of variables.

## 7.5   Conclusions: Spanning Setters and Parallel Architectures

The spanning setter concept developed in this chapter offers a good degree of parallelism with relatively independent tasks on various levels of granularity. With the construction of spanning sets as central mechanism, it is open for an integration of the routes concept, reductions, competition, precision, and modularity on medium and coarse grain levels as well as term parallelism with fine grain size. This makes it a good candidate for an implementation on the different types of parallel architectures discussed here.

On a bus-based system, the implementation can be done with moderate additional effort compared to a sequential one; these system usually provide a global address space and basic mechanisms for the use of shared data structures.

Transputer-based systems, in contrast, only have local memory and must rely on message passing for the exchange of information; experiences with PARTHEO (Schumann and Letz, 1990; Ertel et al., 1989; Bayerl et al., 1989) and parallel unification (Hager and Moser, 1989a) have shown that good exploitation of parallelism is feasible, and that the main obstacles are not the different programming paradigms but lack of experience together with inadequate or malfunctioning program development tools.

Judging the suitability of massively parallel systems for logic programming is difficult since there is almost no experience in that area. The investigations in this chapter show, however, that a concept like spanning setters has a better potential to be successfully mapped onto a massively parallel system than the common parallel execution models for PROLOG.

# Chapter 8

# Conclusions

This chapter resumes the main motives to advocate the use of logic as a tool for parallel programming, as well as the usage of parallel computer systems for the execution of logic programs. During the investigation of various categories of parallelism in logic, it became more and more obvious that logic has the potential to be used for specification purposes as well as for efficient evaluation on parallel computer systems. Being widely accepted as specification mechanism, logic features problems with the control of its evaluation process. To overcome this problem while maintaining good strength as high-level specification mechanism is an essential requirement for an instrument to be used for parallel programming. The concepts presented in the book are critically revised with respect to their capability for solving the above problem.

One of the main issues both in software engineering as well as in hardware design has always been a clear and precise identification of the problem in combination with an accurate description of the envisaged solution, resulting in a more or less formal *specification*. Two important aspects of such a specification are its clarity, leading to an easy comprehension of the problem, and its exactness, mirroring the adequacy of the specification for the problem and its solution. A number of candidates for specification purposes have been in usage, but still an optimal solution has not been found yet. *Natural language* may be convenient for clarity, but is not well suited for checking the correctness. Pure *specification languages* have not seen much acceptance since they require a re-coding for the actual implementation, and then many programmers prefer to directly program the problem. Using a *programming language* as specification tool, or the program itself as its own specification, often is fatal both in terms of understandability and correctness.

A commonly accepted *development methodology* has been stepwise refinement, which is independent of the actual language used. This methodology surely has its advantages for a programmer during the process of developing its program, but again per se satisfies neither clarity nor correctness requirements. In addition to an appropriate programming paradigm which allows the usage of a specification as skeleton for the actual implementation, without re-coding from scratch, and which is open to formal verification, stepwise refinement may come quite close to fulfill both requirements. A good candidate to be mated with stepwise refinement is *logic programming*. It combines a reasonable degree of clarity with a sound formal basis for correctness, and also favors an incremental programming style, adding more and more details to a basic framework.

Parallel programming both as implementing *models* of parallel systems and programming of parallel *computer* systems features two important, complex issues crucial for its usability, efficiency, and acceptance. The first lies in the specification of the problem to be solved. In the sequential case, the behavior of a system to be modeled typically can be mapped onto a single thread of execution[1]; modeling a parallel system requires the management of many threads of execution with a number of interactions, some of them possibly with critical temporal

---

[1] although this may not always be the most natural description

interdependencies. The second is the efficient execution of a program on a parallel computer system, aiming at a performance higher than on a sequential system. The requirement here is to find a mapping of the program structure onto the structure of the underlying hardware which utilizes the available resources in such a way that fast execution is achieved.

The main theme of this book is to advocate the use of logic as a tool for parallel programming while making use of parallel computer systems to speed up the performance of logic programs. It allows the specification of systems to be modeled, or problems to be solved, in a declarative and problem-oriented way, resulting in a concise, comprehensible description open for formal verification. The price to be paid usually is lower efficiency, restricting the use of logic for programming purposes. The reasons for the lack of efficiency are twofold: One is the underlying evaluation mechanism which is more complicated than, e.g., the procedural one, and not as well supported by conventional computers. The second is a lack of control over the evaluation process of a program. Very often the user has some knowledge about an efficient evaluation of his program, but faces problems like missing expressiveness of the language or the need for major modifications of the program. This can be overcome by an integration of 'dirty' constructs (like `cut`, `assert`, `retract` in PROLOG), which greatly depend on the respective evaluation mechanism, to control the evaluation of a program.

This solution obviously is not very satisfying, especially not for the parallel case. An alternative is to describe requirements for an efficient evaluation of a specific program, and model the evaluation process, or maybe even the whole evaluation mechanism, according to these requirements. This solution can be integrated with means of logic by meta-evaluation constructs (which are second-order logic statements about properties of the actual program) where the original program containing the evaluation-independent specification remains unchanged.

Thus, logic can be used both for high-level specification purposes as well as for efficiency-oriented control of the evaluation process. These aspects in general are not solved sufficiently yet, neither by conventional parallel programming languages with a strong control of the execution, nor by high-level languages with good specification properties.

This chapter resumes the main results of the book. After checking the appropriateness of the concepts formulated in the chapter on foundations, the state of the art is compared with the proposals made in the book. Then the potential for the exploitation of parallelism in logic is summarized, and the usability of the parallel logic language MMLOP for the description of parallel systems is discussed. A computational model for this language is resumed with respect to its implementation on different types of parallel architectures.

# 8.1   Foundations

Supporting the use of logic for parallel programming necessitates to clarify some important issues on basic concepts. The calculus used for the evaluation of programs based on logic as well as the language used to write programs surely are one of these, and their usability will be described in the sequel.

### Connection Method

The *connection method* is a sound and complete logic calculus. It can be applied to full first order predicate logic, and is extendible with reasonable effort towards higher order logic, especially for the features required for parallel programming.

### Model Elimination

*Model elimination* can be viewed as a specialization of the connection method. An evaluation of a logic program according to it is based on two proof steps: extension (which is equivalent to input resolution) and reduction (an enhancement for full first order logic). These steps are reasonably powerful, but still can be efficiently automatized and implemented on computers. Their close relationship to the resolution calculus of PROLOG makes an adaptation of Warren's abstract PROLOG machine possible, which is a commonly used execution mechanism for logic programs.

### LOP

The incentive for the development of LOP has been the necessity to create an input language for a family of automated theorem provers. Many interesting problems in addition require further facilities like I/O, which led to an extension of LOP towards a *logic programming language*. It allows the use of full first order predicate logic with various notations, but still is largely compatible to PROLOG by integrating most of its features as a subset. Through a proper integration of global variables and destructive assignment its usability and clarity has been considerably increased, making obsolete 'dirty' constructs like PROLOG's `assert`. Thus its full scope is covered by a clean denotational semantics, while achieving high efficiency in the evaluation through an abstract machine model similar to Warren's one of PROLOG. LOP also in a natural way is open towards higher order logic extensions as those advocated in MMLOP for an adequate treatment of parallel programming.

UNITY

The investigation of the various categories of parallelism has been greatly supported by the use of a (semi-)formal design mechanism for parallel programs. Among the few candidates, UNITY has been chosen as specification language for parallel programs, although at first admittedly mainly for pragmatical reasons, like well-balanced description and numerous examples. It features a clear separation of program design and implementation (coding), aiming at a specification which is independent of the target architecture. It also allows formal reasoning about crucial properties of programs (e.g. termination, reachability of states), and facilitates complexity analysis of parallel programs. Although its computational model relies on state transitions with assignments to variables as basic operation, its use also provided a lot of insight and many valuable hints for the extension of LOP towards parallel programming. It turned out to be appropriate for the specification of parallelism in logic as well as for an investigation of its exploitation. A limitation lies in its restriction to specification purposes, the outcome not being directly usable for implementation. In its present form, it is also not very well suited for the description of complex problems, since it does not incorporate a mechanism like modularity.

## 8.2   State of the Art

The combination of logic programming and parallelism has become more and more popular in the last few years. For this development, two main reasons can be identified. One is the (implicit) exploitation of parallelism in the execution of logic programs aiming at increased performance; the other in addition strives at explicit parallel programming constructs (e.g. for synchronization) while sacrificing the purity of PROLOG's semantics. The first becomes apparent mainly in approaches dealing with OR-parallelism, whereas the second often is connected with AND-parallelism.

## Parallel Logic Languages

Almost all parallel logic languages lack a unified concept for the integration of both *problem-induced* and *evaluation-oriented* parallelism.

AND-parallel languages (also termed committed choice languages) are based on an evaluation mechanism with a different underlying semantics than PROLOG, facilitating the modeling of parallel systems.

OR-parallel languages typically exhibit implicit parallelism emanating from alternative solutions, but maintain the original semantics of PROLOG. OR-parallel languages have problems with the modeling of parallel systems, esp. synchronization (only via facts). Most AND-parallel languages have a different semantics, more oriented towards procedural conceptions. Modeling of parallel systems is achieved via utilization of specific properties of the computational model, e.g. concurrent evaluation of subgoals in the body of a clause to enforce synchronization. Control of the program evaluation is either implicit (through the knowledge of the operational semantics and the evaluation mechanism) or through special constructs, which may require special treatment and destroy the original semantics of a logic program, often in favor of a procedurally oriented evaluation mechanism.

Maybe with the exception of ANDORRA, there is no convincing integration of OR- and AND-parallelism within one single logic language framework as of now.

## Parallel Inference Machines

A number of parallel inference machines have been designed and sometimes built as prototype systems, including implementations on commercially available multiprocessor systems. They are typically designed for a particular AND- or OR-parallel execution model, which is fixed and cannot be adapted easily to differing requirements.

# 8.3 Parallelism in Logic

Most of the attention in the field of parallelism and logic has been paid to AND- and OR-parallelism. These are without doubt the most obvious and probably also the most important ones with respect to increase of performance.

Most of the proposed evaluation mechanisms, however, are based on the dynamic exploitation of AND- or OR-parallelism, and neglect the potential of statical analyses for a better structuring of the execution process.

On the other hand, quite a few different categories of parallelism can be identified which are relevant for logic programming. Their *relevance* not only stems from the ever-present greed for higher performance; parallelism can also be used to facilitate programming, e.g. by program structuring via modules, introduction of abstract data types, or meta-evaluation concepts for a control of the evaluation process[2].

Parallelism in logic programming originates mainly from three *sources*:

- problem-induced: the problem to be solved naturally exhibits some parallelism;
- explicit parallel evaluation: specified by the user;
- implicit parallel evaluation: integrated into the execution mechanism, normally not visible to the user.

Especially the first one very often is neglected or intermingled with the second; in agreement with the separation of specification and implementation, it may prove to be useful in many cases to separate these two clearly.

The *exploitation of parallelism* present in a program also requires some care; it is not sufficient to just identify a lot of potential parallelism without harmonizing its exploitation with the available resources.

One of the central chapters in this book discusses these different categories of parallelism. They have been investigated by identifying the underlying problem and designing a solution for it. Using UNITY as specification language, programs describing the essential features have been derived, which also serve as a basis for the discus-

---

[2] many of these concepts are applicable to or even originate from conventional, sequential programming, but are still more important for the parallel case

sion of important properties, and for refinements. In addition, possible combinations with different categories are outlined. Indications to an implementation are summarized in overview tables on data structures and operations required for computation, storage, and communication purposes.

## 8.4    MMLOP: A Parallel Logic Programming Language

During the investigation of the different categories of parallelism a major problem was an appropriate characterization of their essential features as well as a description suitable for (semi-)formal reasoning about important properties. Although from the very beginning there has been a feeling that logic could serve as basis for a specification language, the simultaneous design of such a language and its application would have been too big a step at once.

In the course of the actual specification process, supported by the use of UNITY, the major characteristics for an extension of LOP gradually became apparent, finally resulting in the proposal of MMLOP. This language enhances the first order predicate logic language LOP by two concepts using second order constructs (predicate variables).

One concept is *modularity* for the expression of parallelism already present in the problem specification.

The second is *meta-evaluation*, which gives the user as much control over the evaluation process as considered necessary.

With these two concepts, MMLOP is appropriate for a reasonably natural description of parallel problems, as well as for an efficient parallel evaluation and execution of logic programs.

### Modularity

The purpose of modularity is to allow structuring of programs into largely independent parts as close as possible to the structure of the problem to be described. This is done in the specification phase, but already embodies the skeleton for the implementation and gives clues for parallel execution: modules are viewed as self-contained units of computation with some communication between them.

For an implementation on a parallel computer system, transformations to map the structure given by the program onto the specific architecture may be necessary; these can be provided by composing the modules in such a way that they form an algebra of modules, and can be subject to formal algebraic transformations. An alternative is to describe such a mapping on a higher level with meta-evaluation constructs. Modularity also provides a smooth way to support high-level programming techniques (abstract data types, libraries).

### Meta-Evaluation

Meta-evaluation designates the description of properties – especially ones important for the evaluation – of an object program in the same language as this object program is given; meta-evaluation can be seen as a combination of meta-programming and partial evaluation. In the context of this book, its main purpose is to allow control of the evaluation and execution of logic programs, esp. on parallel computers. In particular it can be used to adapt the computation rule (evaluation order of subgoals), the search rule (choice of clauses), completeness modes (depth-bound, inference-bound), lemma generation, and the unification mechanism (occur check).

Beyond this more evaluation-oriented usage, meta-constructs are suited for specifications concerning reductions (syntactical transformations to reduce the search space), verification (proving properties of programs, e.g. deadlock-free, termination), theories and inference processors (custom-tailored inference mechanisms), and program development: stepwise refinement, adaptation of the computation mechanism to specific program properties (e.g. Horn / non-Horn, multiple solutions, restriction to propositional logic), debugging, and program maintenance.

Thus, without much knowledge about the actual underlying computational model, the user may tailor an appropriate evaluation mechanism for a particular problem merely by adding some meta-constructs to the original program.

If considered necessary, meta-evaluation can also be applied during the actual execution phase, e.g. for debugging, load balancing, or experimentation with different execution patterns. The use of meta-evaluation at run-time, however, in principle results in self-modifying code which can be difficult to manage.

The combination of modularity and meta-evaluation in MMLOP results in a programming concept with the potential to satisfy most of the requirements imposed onto a specification language for parallel problems, as well as onto an implementation language for the efficient execution of programs on parallel computer systems.

## 8.5   Computational Model

MMLOP basically consists of LOP together with enhancements to deal with modularity and meta-constructs. The strategy of a computational model for the evaluation of MMLOP programs is to transform a MMLOP program into a set of LOP programs plus some communication to be exchanged between them.

This is done in three steps. In the first one, *global meta-evaluation*, all meta-constructs affecting the whole program are eliminated by modifying the object program in such a way that the requirements imposed through these meta-constructs are satisfied. The result is a set of MMLOP programs, each element corresponding to one module.

These programs now are exposed to *modularization*, aiming at a close correspondence between the structure of the whole program as expressed in the modules and their interrelations, and the structure of the target architecture. These two steps have to be performed centrally, typically on the host, since they require knowledge of the whole program structure.

The third step, *local meta-evaluation*, eliminates the meta-constructs local within the modules in analogy to the first step. The resulting programs, which are basically LOP programs enhanced by (built-in) primitives for parallel execution control, are finally executed on the processing elements of the underlying architecture.

This computational model is not very sophisticated yet; it is restricted to a statical application (at compile time) of meta-evaluation and modularization. In principle, however, the concepts can also be applied dynamically at run-time resulting in even greater flexibility of execution control.

In total, the outline of a computational model for efficient execution of MMLOP programs is presented. It is based on transformations of a high-level program into an equivalent one which can be executed more efficiently, allowing some adaptation of the program structure to the target architecture.

# 8.6   Architecture

This section examines the suitability of the MMLOP computational model for different architectures; it is done by discussing the mapping of the spanning setter concept, which is an alternative to model elimination as evaluation mechanism for first order predicate logic formulae, to bus-based, Transputer-based, and massively parallel architectures.[3]

Since communication between modules is required only for problem-specific purposes, both bus-based and Transputer-based architectures are well-suited for an implementation of the spanning setter concept; the first are limited in size by the bus as potential bottleneck while offering a global address space, whereas Transputers provide (at the moment) only four links for point-to-point connections, but can be easily composed into large multi-processor systems. Massively parallel systems, like for example the Connection Machine, require some more effort to implement the spanning setter concept, especially due to their synchronous SIMD mode of operation.

Due to the significance of unification, especially within the spanning setter concept, specific architectures (coprocessors) for unification are investigated in addition. Candidates discussed here are micro-programmable devices, systolic arrays, associative memories, neural networks, and again massively parallel systems (in relatively small instances). Their usage only makes sense in a tight coupling via shared memory (e.g. dual-ported) within the nodes of a parallel architecture.

The spanning setter concept has the advantage that a decomposition of the problem into independent parts (representing alternative solutions) can be performed to a certain degree at compile time; this is very favorable for an implementation on parallel architectures. Its present status of investigation, however, is not sufficient for further conclusions.

As an overall conclusion, logic presents a solid basis for parallel programming. It is usable as a single framework for the specification of parallel problems as well as for implementation purposes on parallel systems. The concepts of modularity and meta-evaluation integrated into the logic programming language MMLOP provide appropriate

---

[3] The model elimination concept and its implementation on parallel architectures as PARTHEO is the subject of a thesis by J. Schumann (Schumann, 1991) and thus not discussed here.

constructs for the description of parallelism and its exploitation during the actual execution of logic programs. The potential of parallelism inherent in logic programming by far exceeds the usually regarded AND-/OR-parallelism; in combination with appropriate computational models – which may be different from PROLOG's resolution calculus – efficient execution of logic programs on parallel architectures can provide sufficient performance to benefit from the excellent descriptive power of logic.

# Acknowledgements

Following the custom used extensively in this book, namely to display information in a structured way using overview tables, the merits of people who made a lot of helpful contributions throughout the evolution of this work are given in the table below. It is arranged in three columns, containing the *symbol*, the *type*, and the *value* of the respective person or institution. Since not even a partial ordering could be induced on the values, the entrances had to be ordered alphabetically with respect to the symbols.

| *symbol* | *type* | *value* |
|---|---|---|
| Baumgartner, P. | colleague | attentive reader, draft version |
| Bayerl, S.; Dr. | project leader ESPRIT 415 F - TUM | advice on logic, typesetting, project management |
| Bibel, W.; Prof. Dr. | founder and leader, Intellektik group | catalyst for my work on logic and parallelism |
| Ertel, W. | project leader ESPRIT 415 - Nixdorf | fruitful project cooperation |
| Fronhöfer, B.; Dr. | colleague | critical reader, draft version |
| Furbach, U.; Dr. | leader of the group | valuable comments on the organization of the book |
| Intellektik group | colleagues | stimulating working atmosphere |
| Jessen, E.; Prof. Dr. | thesis supervisor | superb supervision |
| Kaiser, S. | student & assistant | expansion and maintenance of |
| Kuffer, L. | student & assistant | the bibliographical data base |
| `kimvax` | computer, running a big LaTeX version | broke down only the day *after* the last complicated picture |
| Letz, R. | colleague | discussions logic and parallelism |
| Neugebauer, G. | colleague | LaTeX style connection matrices |
| O'Shaughnessy, C. | private | invaluable |
| Schneeberger, J. | colleague, interim group leader | steered the group through perilous waters |
| Schumann, J. | colleague | discussions parallelism and logic; advice with computer problems |
| Treleaven, P.; Prof. | chairman, WGA, ESPRIT 415 | focal point for issues concerning parallel computer architectures |
| WGA ESPRIT 415 (Working Group on Architecture and Applications) | co-members | exchange of ideas and information about parallel architectures |

# Bibliography


Aho, A. V., Hopcroft, J. E., and Ullman, J. D. (1974). *The Design and Analysis of Computer Algorithms*. Addison-Wesley, Reading, Massachusetts.

Aida, H., Leinwand, S., and Meseguer, J. (1990). Architectural design of the rewrite rule machine ensemble. Technical Report SRI-CSL-90-17, SRI International, Menlo Park, CA 94025.

Aiello, L. and Levi, G. (1984). The Uses of Metaknowledge in AI Systems. In O'Shea, T., editor, *ECAI '88*, pages 705–717. Elsevier.

Albert, B. and Bode, A. (1983). Microprogrammed Associative Instructions: Results and Analysis of a Case Study in Vertical Migrations. In *MICRO-16*.

Ali, K. (1987). OR-parallel execution of Prolog on a multi-sequential machine. *Parallel Programming*, 15(3).

Amthor, R. (1989). Simulation eines Beweisers auf einer Multi-Prozessor Architektur. Master's thesis, Institut für Informatik, Technische Universität München.

Anderson, J., Coates, W., Davis, A., Hon, R., Robinson, I., Robinson, S., and Stevens, K. (1987). The Architecture of FAIM-1. *Computer*, 20:55–67.

Anderson, J. A. and Rosenfeld, E., editors (1988). *Neurocomputing – Foundations of Research*. MIT Press, Cambridge, MA.

Appleby, K., Carlsson, M., Haridi, S., and Sahlin, D. (1988). Garbage Collection for Prolog Based on WAM. *Communications of the ACM*, 31:719–741.

Apt, K. and van Emden, M. (1982). Contributions to the Theory of Logic Programming. *Journal of the Association for Computing Machinery*, 29, No. 3:841–862.

Apt, K. R., Francez, N., and Katz, S. (1988). Appraising fairness in languages for distributed programming. *Distributed Computing 2*, pages 226–241.

Arbab, B. and Berry, D. M. (1987). Operational and Denotational Semantics of Prolog. *Logic Programming*, pages 309–329.

Arbib, M. A. (1989). Schemas and neural networks for sixth generation computing. *Journal of Parallel and Distributed Computing 6*, pages 185–216.

Argonne (1988). *Workshop on Advanced Computer Technologies and Biological Sequencing*. Argonne National Laboratory.

Aristoteles (1976a). *Lehre vom Schluss oder Erste Analytik (Organon III)*. Philosophische Bibliothek. Felix Meiner Verlag, Hamburg.

Aristoteles (1976b). *Lehre vom Beweis oder Zweite Analytik (Organon IV)*. Philosophische Bibliothek. Felix Meiner Verlag, Hamburg.

Aso, M. and Onai, R. (1983). XP's: An Extended OR-Parallel Prolog System. Technical Report TR 023, Institute for New Generation Computer Technology (ICOT).

Aspetsberger, K. (1985). Towards Parallel Machines for Artificial Intelligence: Realization of the ALICE Architecture by the L-Components. In *Austrian Workshop on Artificial Intelligence, Vienna*. Springer.

Attardi, G. and Simi, M. (1984). Metalanguages and Reasoning Across Viewpoints. pages 315–323.

Azema, P., Juanole, G., Sanchis, E., and Montbernard, M. (1984). Specification and Verification of Distributed Systems Using Prolog Interpreted Petri Nets. *IEEE*, pages 510–517.

Bachinger, J. (1987). Implementierung eines parallelen Theorembeweisers und Simulation der Ausführung auf einer Mehrprozessormaschine. Master's thesis, Institut für Informatik, Technische Universität München.

Bal, H. E., Steiner, J. G., and Tanenbaum, A. S. (1989). Programming languages for distributed computing systems. *ACM Computing Surveys*, 21(3):261–322.

Balcke, K. (1987). Spezifikation einer parallelen Version der Konnektions - Methode in der Sprache FP2. Master's thesis, Institut für Informatik, Technische Universität München.

Ballard, D. (1986). Parallel Logical Inference and Energy Minimization. Technical Report TR 142, Computer Science Department, University of Rochester.

Barbuti, R., Levi, G., and Martelli, M. (1987). Modules in Logic Programming: A Survey. Technical report, Dipartimento di Informatica, Universita di Pisa, CNUCE-CNR.

Baron, U., Chassin, J., and Syre, J. (1988). The Parallel ECRC Prolog System PEPSys: An overview and evaluation results. In *FGCS '88*.

Baron, U., Ing, B., Ratcliffe, M., and Robert, P. (1987). A Distributed Architecture for the PEPSys Parallel Logic Programming System. Technical report, ECRC Computer Architecture Group, München.

Bauer, F. L. and Wössner, H. (1981). *Algorithmische Sprache und Programmentwicklung*. Springer, Berlin.

Bayerl, S., Letz, R., Kurfeß, F., Schumann, J., and Ertel, W. (1989). PARTHEO: Full First Order Logic Parallel Inference Machine – Language and Design –. Deliverable D16, ESPRIT 415 F, Institut für Informatik, Technische Universität, München.

Beer, J. (1989). *Concepts, Design, and Performance Analysis of a Parallel Prolog Machine*, volume 404 of *Lecture Notes in Computer Science*. Springer.

Beer, J. and Giloi, W. K. (1987). POPE - A Parallel-Operating Prolog Engine. *Future Generations Computer Systems*, pages 83–92.

Bellia, M. and Levi, G. (1986). The relation between logic and functional languages: A survey. *J. Logic Programming*, 3:217–236.

Ben-Ari, M. (1984). *Principles of Concurrent Programming*. Prentice Hall.

Bibel, W. (1980). On Matrices with Connections. *Journal of the ACM*, 28:633–645.

Bibel, W. (1983). Matings in Matrices. *Communications of the ACM*, 26, Nr. 11:844–852.

Bibel, W. (1986). Methods of Automated Reasoning. In (Bibel and Jorrand, 1986), pages 171–217.

Bibel, W. (1987). *Automated Theorem Proving*. Vieweg, Braunschweig, Wiesbaden, second edition.

Bibel, W. (1988). Advanced topics in automated deduction. In Nossum, R., editor, *Fundamentals of Artificial Intelligence II*, Berlin. Springer.

Bibel, W. (1990). Perspectives of Automated Deduction. In (Stickel, 1990b).

Bibel, W. and Aspetsberger, K. (1985). A Bibliography on Parallel Inference Machines. *Symbolic Computation*, 1(1):115–118.

Bibel, W. and Buchberger, B. (1984). Towards a Connection Machine for Logic Inference. *Future Generation Computer Systems*, 1(3):177–188.

Bibel, W. and Jorrand, P., editors (1986). *Fundamentals of Artificial Intelligence*, volume 232 of *Lecture Notes in Computer Science*, Berlin. Springer.

Bibel, W., Kurfeß, F., Aspetsberger, K., Hintenaus, P., and Schumann, J. (1987). Parallel inference machines. In (Treleaven and Vanneschi, 1987), pages 185–226.

Bic, L. (1984). A Data-Driven Model for Parallel Interpretation of Logic Programs. In *Proceedings of the International Conference on Fifth Generation Computer Systems 1984*, pages 517–523. Institute for New Generation Computer Technology (ICOT).

Bläsius, K. H. and Bürckert, H. J. (1987). *Deduktionssysteme*. Oldenbourg, München.

Bobrow, D. (1984). If Prolog is the Answer, what is the Question? In *International Conference on Fifth Generation Computer Systems*, pages 138–145. Institute for New Generation Computer Technology (ICOT).

Böck, K.-H. (1989). Studying an application for a parallel logic programming system. Master's thesis, Institut für Informatik, Technische Universität München.

Bode, A. (1980). Vertical Processing: The Emulation of Associative and Parallel Behavior on Conventional Hardware. *Microprocessor Systems*, pages 215–220.

Boole, G. (1854). An Investigation of the Laws of Thought, on which are founded the Mathematical Theory of Logic and Probabilities. London.

Borgwardt, P. (1984). Parallel Prolog Using Stack Segments on Shared Memory Multiprocessors. In *International Symposium On Logic Programming*, Atlantic City, NJ.

Bowen, K. and Kowalski, R. (1982). Amalgamating Language and Metalanguage in Logic Programming. *Logic Programming*.

Boyer, R. and Moore, J. (1979). *A Computational Logic*. Academic Press, New York.

Boyer, T. and Moore, J. (1983). A Mechanical Proof of the Turing Completeness of Pure LISP. Technical report, Institute for Computer Science, University of Texas at Austin.

Brandes, T. (1988). *Formale Methoden zur Spezifizierung automatischer Parallelisierung*. Hochschultexte Informatik. Dr. Alfred Hüthig Verlag, Heidelberg.

Bratko, I. (1986). *Prolog Programming for Artificial Intelligence*. Addison-Wesley.

Brauer, W. and Freksa, C., editors (1989). *Wissensbasierte Systeme 89*, Berlin. Springer.

Brogi, A. and Gorrieri, R. (1989). A Distributed, Net Oriented Semantics for Delta Prolog. In *TAPSOFT '89*, pages 162–177.

Bruynooghe, M. and Pereira, L. M. (1984). Deduction revision by intelligent backtracking. In Campbell, J. A., editor, *Implementations of PROLOG*, pages 194–215. Horwood, Chichester, England.

Caferra, R. and Jorrand, P. (1985). Unification with Refined Linearity Check as a Network of Parallell Processes. Technical report, LIFIA, Laboratoire d'Informatique Fondamentale et d'Intelligence Artificielle IMAG, Grenoble, France.

Carriero, N. and Gelernter, D. (1989a). How to write parallel programs: A guide to the perplexed. *ACM Computing Surveys*, 21(3):323–357.

Carriero, N. and Gelernter, D. (1989b). Linda in context. *Communications of the ACM*, 32(4):444–458.

Chandy, K. M. and Misra, J. (1988). *Parallel Program Design*. Addison-Wesley, Reading, MA.

Chang, C. and Lee, R. (1973). *Symbolic Logic and Mechanical Theorem Proving*. Academic Press, New York.

Chang, J.-H. (1985). High Performance Execution of Prolog Programs Based on A Static Data Dependency Analysis. Technical Report UCB/CSD 86/263, University of California, Berkeley.

Chang, J.-H. and Despain, A. (1986). Semi-Intelligent Backtracking of Prolog: Based on A Static Data Dependency Analysis. In *Logic Programming Conference*, Berkeley.

Chassin de Kergommeaux, J., Baron, U., Rapp, W., and Ratcliffe, M. (1988). Performance Analysis of Parallel Prolog: A Correlated Approach. Technical report, ECRC Munich.

Chassin de Kergommeaux, J., Codognet, P., Robert, P., and Syre, J.-C. (1989). Une programmation logique parallele: premiere partie: Langages gardes. *Technique et Science Informatiques*, 8:205–224.

Chen, C., Singhal, A., and Patt, Y. N. (1988). PUP: An Architecture to Exploit Parallel Unification in Prolog. Technical Report UCB/CSD 88/414, University of California, Computer Science Department), Berkeley, CA 94720.

Chenadec, P. L. (1989). On the logic of unification. *Journal of Symbolic Computation*, (8):141–199.

Chu, Y. and Itano, K. (1984). Organisation of a Parallel PROLOG Machine. *Proc. Intern. Workshop on HLCA*.

Church, A. (1956). *Introduction to Mathematical Logic*. Princeton University Press, Princeton.

Ciepielewski, A. and Haridi, S. (1984a). Control of Activities in the OR-parallel Token Machine. Technical report, Department Of Telecommunications and Comping Systems, Royal Institute of Technology, Stockholm.

Ciepielewski, A. and Haridi, S. (1984b). Execution of Bagof on the OR-parallel Token Machine. In *International Conference on Fifth Generation Computer Systems*, pages 551–562, Tokyo.

Citrin, W. (1985). A Comparison of Indexing and Term Reordering: Two Methods
for Speeding Up the Execution of Prolog Programs. Technical Report CS 257,
University of California, Berkeley.

Citrin, W. (1988). Parallel Unification Scheduling in Prolog. Technical Report
UCB/CSD 88/415, Department of Electrical Engineering and Computer Science, University of California, Berkeley, CA.

Citrin, W., van Roy, P., and Despain, A. (1986). Compiling Prolog for the Berkeley
PLM. In *19th Annual Hawaii Intern. Confernce on System Science*.

Civera, P., Piccinini, G., and Zamboni, M. (1989). Implementation Studies for a
VLSI Prolog Coprocessor. *IEEE Micro*, pages 10–23.

Clark, K. (1984, revised June 1985). Parlog: Parallel Programming in Logic.
Technical report.

Clark, K. (1988). Parlog and Its Applications. *IEEE Transactions on Software
Engineering*, 14:1792–1804.

Clark, K. and Gregory, S. (1981). A Relational Language for Parallel Programming. In *ACM Conference on Functional Programming Languages and Computer Architecture*, pages 171–178.

Clark, K. and Gregory, S. (1984). Notes on Systems Programming in Parlog. *Proc.
Of the International Conference On Fifth Generation Computer Systems*,
pages 299–306.

Clark, K. and Gregory, S. (1986). PARLOG: Parallel Programming in Logic. *ACM
Transactions on Programming Languages and Systems*, 1986(8):1–49.

Clocksin, W. (1985). Design and Implementation of a Sequential PROLOG Machine. *New Generation Computing*, 3(1):101–119.

Clocksin, W. (1987). A Prolog Primer. *BYTE*, (8):147–158.

Clocksin, W. and Alshawi, H. (1988). A Method for Efficiently Executing Horn
Clause Programs Using Multiple Processors. *New Generation Computing*,
(5):361–376.

Clocksin, W. and Mellish, C. (1984). *Programming in Prolog*. Springer.

Cohen, J. (1988). A View of the Origins and Development of Prolog. *Communications of the ACM*, 31:26–36.

Colmerauer, A. (1982). Prolog and Infinite Trees. *Logic Programming*, pages
231–251.

Colmerauer, A. (1987). Opening the Prolog III Universe. *BYTE*, 8:177–182.

Colmerauer, A., Pasero, H., and Roussel, P. (1973). Un systeme de communication
homme-machine en Francais. Technical report, Artificial Intelligence Group,
University of Aix-Marseilles, Luminy, France.

Comon, H. (1988). *Unification et disunification. Théorie et applications*. PhD
thesis, I.N.P. de Grenoble, France.

Comon, H. and Lescanne, P. (1989). Equational problems and disunification.
*Journal of Symbolic Computation*, 7:371–425.

Conery, J. S. (1983). *The AND/OR Process Model for Parallel Execution of Logic
Programs*. PhD thesis, University of California, Irvine. Technical report 204,
Department of Information and Computer Science.

Cook, S. and Reckhow, R. (1974). On the Lengths of Proofs in the Propositional
Calculus. In *ACM Symposium on Theory of Computing '74*, pages 135–148.

Corbin, J. and Bidoit, M. (1983). A Rehabilitation of Robinson's Unification Algorithm. *Information Processing '83*, pages 909–914.

Corsini, P., Frosini, G., and Rizzo, L. (1989). Implementing a Parallel PRO-LOG Interpreter by Using OCCAM and Transputers. *Microprocessors and Microsystems*, 13(4):271–279.

Corsini, P., Frosini, G., and Speranza, G. (1989). The Parallel Interpretation of Logic Programs in Distributed Architectures. *Computer Journal*, 32:29–35.

Crammond, J. (1985). A Comparative Study of Unification Algorithms for OR-Parallel Execution of Logic Languages. *IEEE*, pages 131–138.

Crammond, J. (1986). An Execution Model for Committed-Choice Nondeterministic Languages. In *Symposium on Logic Programming '86*, pages 148–158.

Cunha, J. C., Ferreira, M. C., and Moniz Pereira, L. (1989). Programming in Delta Prolog. In *Logic Programming Conference '89*.

Dai, K. (1988). *Large-Grain Dataflow Computation and its Architectural Support*. PhD thesis, Technische Universität Berlin, Fachbereich 20 Informatik.

Damm, W. and Doehmen, G. (1989). Specifying distributed computer architectures in AADL. *Parallel Computing*, (9):193–211.

Darlington, J., Field, A., and Pull, H. (1985). The Unification of Functional and Logic Languages. Technical report, Imperial College of Science and Technology, Department of Computing, London.

Darlington, J. and Reeve, M. (1981). ALICE - A Multi-Processor Reduction Machine for the Parallel Evaluation of Applicative Languages. In *ACM Conference on Functional Programming Languages and Computer Architecture*, pages 65–74.

Darlington, J. and Reeve, M. (1983). ALICE and the Parallel Evaluation of Logic Programs. *10th Annual International Symposium on Computer Architecture*.

Davis, A. and Robison, S. The Architecture of the FAIM-1 Symbolic Multiprocessing System. Technical report, Artificial Intelligence Laboratory, Schlumberger Palo Alto Research, Palo Alto, CA.

De Bakker, J., De Roever, W., and Rozenberg, G., editors (1986). *Current Trends in Concurrency*, volume 224 of *Lecture Notes in Computer Science*. Springer, Berlin.

De Morgan (1847). Formal Logic, or The Calculus of Inference, Necessary and Probable. London.

Deering, M. F. (1984). Hardware and Software Architectures for Efficient AI. In *National Conference on Artificial Intelligence (AAAI)*, pages 73–78.

DeGroot, D. (1984). Restricted AND-Parallelism. In *International Conference on Fifth Generation Computer Systems*, pages 471–478. Institute for New Generation Computer Technology (ICOT).

DeGroot, D. (1988). A Technique for Compiling Execution Graph Expressions for Restricted And-Parallelism in Logic Programs. *Journal of Parallel and Distributed Computing*, (5):494–516.

Delcher, A. L. and Kasif, S. (1988). Efficient Parallel Term Matching. Technical report, Computer Science Department, Johns Hopkins University, Baltimore, MD 21218.

Denning, P. (1986). Parallel Computing and Its Evolution. *Communications of the ACM*, 29:1163–1169.

Despain, A. and Patt, Y. (1984). The Aquarius Project. In *COMPCON '84*, Berkeley, CA. University Of California.

Despain, A. M. and Patt, Y. N. (1985). Aquarius - A High Performance Computing System for Symbolic and Numeric Applications. In *COMPCON '85*, Berkeley, CA.

Diel, H., Lenz, N., and Welsch, H. M. (1986). System Structure for Parallel Logic Programming. *Future Generation Computer Systems*, 2:225–231.

Dijkstra, E. (1976). *A Discipline of Programming*. Prentice Hall, Englewood Cliffs, NJ.

Dilger, W. and Janson, A. (1986). Intelligent Backtracking in Deduction Systems by Means of Extended Unification Graphs. *Journal of Automated Reasoning*, 2:43–62.

Dilger, W. and Müller, J. (1984). Punify - An AI-Machine for Unification. In *International Conference on Fifth Generation Computer Systems*, page 615ff. Elsevier.

Dilger, W. and Schneider, H.-A. ASSIP-T. A Theorem Proving Machine. Technical report, Fraunhofer-Institut für Informations- und Datenverarbeitung Karlsruhe.

Dincbas, M. and Pape, J.-P. L. (1984). Metacontrol of Logic Programs in META-LOG. In *International Conference on Fifth Generation Computer Systems*, pages 361–370.

Dobry, T., Despain, A., and Patt, Y. (1985). Performance Studies of a Prolog Machine Architecture. In *12th International Symposium on Computer Architecture*.

Dolan, C. P. (1989). Tensor Manipulation Networks: Connectionist and Symbolic Approaches to Comprehension, Learning, and Planning. Technical Report UCLA-AI-89-06, University of California, Los Angeles.

Dörfler, W. and Mühlbacher, J. (1973). *Graphentheorie für Informatiker*. de Gruyter, Berlin.

Dreyfus, H. L. and Dreyfus, S. E. (1988). Making a Mind Versus Modeling the Brain: Artificial Intelligence Back at a Branchpoint. *Daedalus*, 117 (Winter 1988):15–43.

Dwork, C., Kanellakis, P., and Stockmeyer, L. (1986). Parallel Algorithms for Term Matching. In *CADE '86*, pages 416–430, Berlin. Springer.

Dwork, C., Kanellakis, P. C., and Stockmeyer, L. (1988). Parallel algorithms for term matching. *SIAM Journal of Computing 4*, 17(4):711–731. Unification, term matching, parallel algorithms, logic programming.

Eckmiller, R., Hartmann, G., and Hauske, G., editors (1990). *Parallel Processing in Neural Systems and Computers*. Elsevier.

Eder, E. (1985a). An Implementation of a Theorem Prover Based on the Connection Method. Technical report, Institut für Informatik, Technische Universität München.

Eder, E. (1985b). Properties of Substitutions and Unifications. *Journal of Symbolic Computation*, (1):31–46.

Eder, E. (1989). Relative Complexities of First Order Calculi. Habilitation thesis, Lehrstuhl für Informatik, Universität Dortmund. to appear in Vieweg Verlag, Wiesbaden.

Eisenstadt, M. and Brayshaw, M. (1988). The Transparent Prolog Machine (TPM): An Execution Model and Graphical Debugger for Logic Programming. *Logic Programming*, pages 277–342.

Enderton, H. and Kleene, S. (1967). *Mathematical Logic*. John Wiley and Sons, New York.

Engels, J. (1988). A Model for Or-parallel Execution of (Full) Prolog and its Proposed Implementation. Technical report, Institut f. Informatik III, University Bonn, Bonn.

Eppler, W. (1990). Implementation of fuzzy production systems with neural networks. In (Eckmiller et al., 1990), pages 249–252.

Ertel, W. (1988). Backpropagation with temperature parameter and random pattern presentation. Technical report, Institut für Informatik, Technische Universität München.

Ertel, W., Schumann, J., Letz, R., Bayerl, S., Kurfeß, F., van der Koelen, M., Suttner, C., and Trapp, N. (1989). PARTHEO: Parallel Automated Theorem Prover based on the Connection Method for Full First Order Logic. Deliverable D15, ESPRIT 415 F, Institut für Informatik, Technische Universität, München.

Fagin, B. and Despain, A. M. (1987). Performance Studies of a Parallel Prolog Architecture. Technical report, Computer Science Division, University of California, Berkeley, CA. preprint from ISCA, June '87.

Fagin, B. S. and Despain, A. M. (1990). The performance of parallel Prolog programs. *IEEE Transactions on Computers*, 39(12):1434–1445.

Fahlman, S. and Hinton, G. (1987). Connectionist Architectures for Artificial Intelligence. *Computer*, 20:100–118.

Feldman, J. A., Fanty, M. A., Goddard, N. H., and Lynne, K. J. (1988). Computing with Structured Connectionist Networks. *Communications of the ACM*, 31:170–187.

Findler, N. and Gao, J. (1987). Dynamic hierachical control for distributed problem solving. *Data & Knowledge Engineering*, 2:285–301.

Finkel, R. and Manber, U. (1987). DIB - A Distributed Implementation of Backtracking. *ACM Transactions on Programming Languages and Systems*, 9:235–256.

Fodor, J. A. (1983). *The Modularity of Mind*. MIT Press, Cambridge, MA.

Foster, I. (1990). *Systems Programming in Parallel Logic Languages*. Prentice Hall.

Foster, I. and Taylor, S. (1987). Flat Parlog: A Basis for Comparison. *Parallel Programming*, 16:87–125.

Foster, I. and Taylor, S. (1989). *STRAND: New Concepts in Parallel Programming*. Prentice Hall.

Fox, G. C. and Messina, P. C. (1987). Advanced Computer Architectures. *Scientific American*, 257:45–52.

Frege, G. (1879). *Begriffsschrift, eine der arithmetischen nachgebildete Formelsprache des reinen Denkens*. Nebert Verlag.

Freisleben, B. (1987). *Mechanismen zur Synchronisation paralleler Prozesse*. Number 133 in Informatik-Fachberichte. Springer, Berlin.

Frenkel, K. A. (1986). Evaluating Two Massively Parallel Machines. *Communications of the ACM*, 29:752–758.

Futamura, Y. (1988). Program Evaluation and Generalized Partial Computation. In *International Conference on Fifth Generation Computer Systems*, pages 685–692, Tokyo. Institute for New Generation Computer Technology (ICOT).

Futo, I. (1988). Parallele Programmierung in CS-Prolog. *Artificial Intelligence Newsletter*, 9, 10:13–15, 16–19.

Futo, I. and Kacsuk, P. (1989). CS-Prolog on multitransputer systems. *Microprocessors and Microsystems*, 13:103–112.

Gallaire, H. (1986). Merging Objects and Logic Programming: Relational Semantics. In *AAAI '86*, pages 755–757.

Gaudiot, J., Pi, J., and Campbell, M. (1988). Program graph allocation in distributed multicomputers. *Parallel Computing*, 7:227–247.

Gelernter, D. (1987). Programming for Advanced Computing. *Scientific American*, 257:65–71.

Gelernter, D. (1988). Getting the Job Done. *BYTE*, (11):301–307.

Genesereth, M. and Ginsberg, M. (1985). Logic Programming. *Communications of the ACM*, 28(9):933–941.

Gentzen, G. (1935). Untersuchungen über das logische Schließen. *Mathematische Zeitschrift*, 39:176–210, 405–431.

Georgescu, I. (1986). An Inference Processor based on reactive memory. Technical report, Institute for Computers and Informatics, Department of Robotics and Artificial Intelligence, Bucharest.

Ghosh, J. and Hwang, K. (1989). Mapping neural networks onto message-passing multicomputers. *Journal of Parallel and Distributed Computing 6*, pages 291–330.

Giambiasi, N., Lbath, R., and Touzet, C. (1989). Une approche connexionniste pour calculer l'implication floue dans les systemes a base de regles. Technical report, Universite de Nimes.

Giloi, W. K. (1981). *Rechnerarchitektur*. Springer, Berlin.

Goguen, J. A. (1988). Higher Order Functions Considered Unnecessary for Higher Order Programming. Technical report, SRI International, Menlo Park, CA 94025.

Goguen, J. A. (1990). Semantic specifications for the rewrite rule machine. Technical Report SRI-CSL-90-13, SRI International, Menlo Park, CA 94025.

Goguen, J. A., Meseguer, J., Leinwand, S., Winkler, T., and Aida, H. (1989). The Rewrite Rule Machine Project. Technical report, SRI International.

Goguen, J. A. and Meseguer, J. (1987). Unifying Functional, Object-Oriented and Relational Programming with Logical Semantics. Technical report, SRI International, Menlo Park, CA 94025.

Gonauser, M. and Mrva, M. (1989). *Multiprozessor-Systeme: Architektur und Leistungsbewertung*. Springer, Berlin.

Goto, A., Aida, H., Maruyama, T., Yuhara, M., Tanaka, H., and Moto-OKA, T. (1983). A Highly Parallel Inference Engine: PIE. In *Logic Programming Conference*. Institute for New Generation Computer Technology (ICOT).

Goto, A., Sato, M., Nakajima, K., Taki, K., and Matsumoto, A. (1988). Overview of the Parallel Inference Machine Architecture (PIM). In *International Conference on Fifth Generation Computer Systems*, pages 209–229, Tokyo. Institute for New Generation Computer Technology (ICOT).

Goto, A., Tanaka, H., and Moto-Oka, T. (1984). Highly Parallel Inference Engine: PIE. Goal Rewriting Model and Machine Architecture. *New Generation Computing*.

Gregory, S. (1984). Implementing PARLOG on the ALICE Machine. Technical report, Imperial College, London.

Gregory, S., Foster, I. T., Burt, A. D., and Ringwood, G. A. (1989). An Abstract Machine for the Implementation of PARLOG on Uniprocessors. *New Generation Computing*, (6):389–420.

Güntzer, U., Kiessling, W., and Bayer, R. (1986). Evaluation Paradigms for Deductive Databases: from Systolic to As-You-Please. Technical Report TUM-I-86-05, Institut für Informatik, Technische Universität München.

Hager, J. and Moser, M. (1989a). An Approach to Parallel Unification Using Transputers. In *GWAI '89*, pages 83–91. Springer.

Hager, J. and Moser, M. (1989b). PASTE - A Parallel System for Term Unification. Fortgeschrittenenpraktikum, Institut für Informatik, Technische Universität München.

Hailperin, M. and Westphal, H. (1986). A Computational Model for PEPSys. Technical Report CA-16, ECRC.

Halim, Z. (1986). A Data-Driven Machine for Or-Parallel Evaluation of Logic Programs. *New Generation Computing*, Vol.4, No.1:5–33.

Halpern, J. Y. (1990). An analysis of first-order logics of probability. *Artificial Intelligence*, 46:311–350.

Händler, W. (1984). Computer Architecture and Applications: Complexity and Flexibility. *Computers and Artificial Intelligence*, 3:79–104.

Haridi, S. and Brand, P. (1988). ANDORRA Prolog – An Integration of Prolog and Commited Choice languages. In *Future Generation Computer Systems*, pages 745–754, Tokyo. Institute for New Generation Computer Technology (ICOT).

Haridi, S. and Ciepielewski, S. (1983). An OR-parallel Token Machine. In *Logic Programming Workshop '83*, Praio da Falesia, Portugal.

Harland, J. and Jaffar, J. (1987). On Parallel Unification for Prolog. *New Generation Computing*, 5:259–279.

Hattori, A., Shinogoi, T., Kumon, K., and Goto, A. (1989). PIM-p: A Hierarchical Parallel Inference Machine. Technical Report TR-514, Institute for New Generation Computer Technology (ICOT), Tokyo, Japan.

Haugeland, J., editor (1981). *Mind Design: Philosophy, Psychology, Artificial Intelligence*. MIT Press, Cambride, MA.

Hayes, J., Michie, D., and Pao, Y.-H. (1982). *Machine Intelligence 10*. Ellis Horwood Limited, John Wiley & Sons, New York, Brisbane, Chichester, Toronto.

Hellerstein, L. and Shapiro, E. (1986). Implementing Parallel Allgorithms in Concurrent Prolog: The Maxflow Experience. *Logic Programming*, 2:157–184.

Hennessy, J. L. and Patterson, D. A. (1990). *Computer Architecture: A Quantitative Approach*. Morgan Kaufmann, San Mateo, CA.

Hentenryck, P. and Dincbas, M. (1986). Domains in Logic Programming. In *AAAI '86*, pages 759–765.

Herbrand, J. (1930). Recherches sur la theorie de la demonstration. Universite de Paris. In: Ecrits logiques de Jaques Herbrand. Paris, PUF, 1968.

Hermenegildo, M. (1986). An Abstract Machine for Restricted AND-Parallel Execution of Logic Programs. In *Third International Conference On Logic Programming 86*, pages 25–39.

Hermenegildo, M. and Nasr, R. (1986). Efficient Management of Backtracking in AND-Parallelism. In *Third International Conference On Logic Programming 86*, pages 40–54.

Hermes, H. (1976). *Einführung in die mathematische Logik*. B.G. Teubner, Stuttgart.

Hermes, H. (1978). *Aufzählbarkeit, Entscheidbarkeit, Berechenbarkeit*. Springer, Berlin.

Hertzberger, L. and van de Riet, R. (1984). Progress in the Fith Generation Inference Architectures. *Future Generations Computer Systems*, 1(2):93–102.

Hierata, K. and et al. (1983). An Efficient Processing Method of Structured Data on the Highly Parallel Inference Engine PIE. Technical report, Japan.

Hilbert, D. and Ackermann, W. (1950). *Principles of Mathematical Logic*. Chelsea Pub. Co., New York.

Hillis, D. W. (1985). *The Connection Machine*. MIT Press, Cambridge, MA.

Hillis, W. and Steele, G. (1986). Data Parallel Algorithms. *Communications of the ACM*, 29:1170–1183.

Hinton, G. E. (1984). Distributed Representations. Technical report, Computer Science Department, Carnegie-Mellon University, Pittsburgh, PA 15213.

Hoare, C. (1985). *Communicating Sequential Processes*. Prentice Hall.

Hofstadter, R. (1979). *Gödel, Escher, Bach: An Eternal Golden Braid*. Basic Books, New York.

Hölldobler, S. (1990a). CHCL – a connectionist inference system for Horn logic based on the connection method. Technical Report TR-90-042, International Computer Science Institute, Berkeley, CA 94704.

Hölldobler, S. (1990b). A connectionist unification algorithm. Technical Report TR-90-012, International Computer Science Institute, Berkeley, CA 94704.

Hölldobler, S. (1990c). A structured connectionist unification algorithm. In *AAAI '90*, pages 587–593.

Hölldobler, S. (1990d). Towards a connectionist inference system. In *Proceedings of the International Symposium on Computational Intelligence*.

Honavar, V. and Uhr, L. (1990). Symbol processing systems, connectionist networks, and generalized connectionist networks. Technical Report 90-23, Department of Computer Science, Iowa State University, Ames, IA 50011.

Horning, J. and Randell, B. (1973). Process Structuring. *ACM Computing Surveys*, 5:5–30.

Hudak, P. (1989). Conception, evolution, and application of functional programming languages. *ACM Computing Surveys*, 21(3):359–411.

Huet, G. (1986). Deduction and Computation. In (Bibel and Jorrand, 1986), pages 39–74.

Hufflen, J. (1989). *Fonctions et généricité dans un langage de programmation parallèle.* PhD thesis, I.N.P. de Grenoble, France.

Hwang, K. and Briggs, F. (1984). *Computer Architecture and Parallel Processing.* Mc Graw-Hill, New York.

Ibañez, M. B. (1988). Parallel inferencing in first-order logic based on the connection method. In *Artificial Intelligence: Methodology, Systems, Applications '88.* Varna, North-Holland.

Ibañez, M. B. (1989). *Inférence parallèle et processus communicants pour les clauses de Horn. Extension au premier ordre par la méthode de connexion.* PhD thesis, I.N.P. de Grenoble, France.

ICOT (1984). Outline of fifth generation computer project. Technical report, Institute for New Generation Computer Technology (ICOT), Tokyo, Japan.

INMOS (1986). *Transputer Architecture.* INMOS Ltd.

INMOS (1988a). *Occam 2 Reference Manual.* Prentice Hall.

INMOS (1988b). *Transputer Databook.* INMOS Ltd.

Ino, E. and Koelbl, D. (1988). Sequentielle und parallele Architekturansätze für logische Programmiersprachen. *Informatik Forschung und Entwicklung*, 3:182–194.

Itai, A. and Makowsky, J. (1983). Unification as a complexity measure for logic programming. Technical report, Technion, Israel Institute of Technology, Department of Computer Science, Haifa, Israel.

Ito, N., Kuno, E., and Oohara, T. (1987). Efficient Stream Processing in GHC and Its Evaluation on a Parallel Inference Machine. *Journal of Information Processing*, 10:237–244.

Ito, N., Onai, R., Masuda, K., and Shimizu, H. (1983). Parallel Prolog Machine Based on Data Flow Mechanism. In *Logic Programming Conference '83.* Institute for New Generation Computer Technology (ICOT).

Jackoway, G. (1984). Associative Networks on a Massively Parallel Computer. Technical report, Department of Computer Science, Duke University.

Jaffar, J. (1984). Efficient Unification over Infinite Terms. *New Generation Computing*, 2:207–219.

Jones, N., Asestoft, P., and Sondergaard, H. (1985). An Experiment in Partial Evaluation: The Generation of a Compiler Generator. In *First International Conference On Rewriting Techniques and Applications*, Dijon, France.

Jorrand, P. (1986). Term Rewriting as a Basis for the Design of a Functional and Parallel Programming Language. A case study: the Language FP2. In (Bibel and Jorrand, 1986), pages 221–276.

Jorrand, P. (1987). Design and Implementation of a Parallel Inference Machine for First-Order Logic: An Overview. In *PARLE 87*, volume 258 of *Lecture Notes in Computer Science*, Berlin. Springer.

Kahn, K. e. a. (1986). Objects in concurrent logic languages. In *OOPSLA 86*, pages 242–257.

Kaindl, H. (1989). *Problemlösen durch heuristische Suche in der Artificial Intelligence.* Springer, Wien.

Kalé, L. V. (1988). A Tree Representation for Parallel Problem Solving. In *AAAI '88*, pages 677–681.

Kalé, L. V. (1989). The reduce OR process model for parallel execution of logic programs. *Journal of Logic Programming*.

Kalé, L. V. and Ramkumar, B. (1990). Joining and parallel solutions in AND/OR parallel systems: Part I - static analysis. In *ICLP '90*.

Kasif, S., Kohli, M., and Minker, J. (1983). PRISM: A Parallel Inference System for Problem Solving. In *Logic Programming Workshop '83*, pages 123–152, Lisboa, Portugal. Universidade Nova de Lisboa.

Kimura, Y. and Chikayama, T. (1987). An Abstract KL1 Machine and its Instruction Set. Technical Report TR-246, Institute for New Generation Computer Technology (ICOT), Tokyo, Japan.

Kleinknecht, R. and Wüst, E. (1976a). *Lehrbuch der elementaren Logik Band 1: Aussagenlogik*, volume 4118 of *Wissenschaftliche Reihe*. dtv, München.

Kleinknecht, R. and Wüst, E. (1976b). *Lehrbuch der elementaren Logik Band 2: Prädikatenlogik*, volume 4119 of *Wissenschaftliche Reihe*. dtv, München.

Kliger, S., Yardeni, E., Kahn, K., and Shapiro, E. (1988). The Language FCP(:,?). In *FGCS '88*, pages 763–783, Tokyo. Institute for New Generation Computer Technology (ICOT).

Kneale, W. and Kneale, M. (1984). *The Development of Logic*. Clarendon Press, Oxford.

Knight, K. (1989). Unification: A multidisciplinary survey. *ACM Computing Surveys*, 21(1):93–124.

Knuth, D. (1968). *The Art of Computer Programming: Volume 1, Fundamental Algorithms*. Addison-Wesley, Reading, MA.

Knuth, D. (1973). *The Art of Computer Programming: Volume 3, Sorting and Searching*. Addison-Wesley, Reading, MA.

Kober, R., editor (1988). *Parallelrechner-Architekturen*. Springer, Berlin.

Kohli, M. and Minker, J. Control of Logic Programs Using Integrity Constraints. Technical report, Department of Computer Science, University of Maryland.

Kohonen, T. (1987). *Content-Addressable Memories*. Springer, Berlin.

Komiya, S. (1989). Automatic programming by composing program components and its realization method. *Future Generation Computer Systems 5*, pages 151–161.

Kowalski, R. (1979). Algorithm = logic + control. *Communications of the ACM*, 22:424–431.

Kowalski, R. (1983). Logic programming. *Information Processing*, pages 133–145.

Kowalski, R. (1984). The history of logic programming. *LP Research Reports*.

Kowalski, R. (1985). The Limitations of Logic. Technical report, Department of Computing, Imperial College, University of London.

Kowalski, R. (1988). The Early Years of Logic Programming. *Communications of the ACM*, 31:38–43.

Kowalski, R. and Sergot, M. (1986). A Logic-based Calculus of Events. *New Generation Computing*, 4(1):67–95.

Kronsjo, L. (1985). *Computational Complexity of Sequential and Parallel Algorithms*. Wiley.

Kung, C.-H. (1985). High Parallelism and a Proof Procedure. *Decision Support Systems*, 1:323–331.

Kurfeß, F. (1991). Unification on a connectionist simulator. In *International Conference on Artificial Neural Networks – ICANN-91*, Helsinki, Finland.

Kurfeß, F., Pandolfi, X., Belmesk, Z., Ertel, W., Letz, R., and Schumann, J. (1989). PARTHEO and FP2: Design of a parallel inference machine. In Treleaven, P., editor, *Parallel Computers: Object-Oriented, Functional and Logical*, chapter 9. Wiley.

Kurfeß, F. and Reich, M. (1989). Logic and reasoning with neural models. In *Connectionism in Perspective*, pages 365–376, Amsterdam. Elsevier.

Kurfeß, F. and Schumann, J. (1989). MetaLOP – Basic Concepts and Ideas. Technical report, Institut für Informatik, Technische Universität, München. Draft.

Kurozumi, T. (1989). Outline of the fifth generation computer systems project and ICOT activities. Technical Report TR-523, Institute for New Generation Computer Technology (ICOT), Tokyo, Japan.

Kursawe, P. (1986). How to Invent a Prolog Machine. In *Third International Conference On Logic Programming '86*, pages 135–148.

Lamport, L. (1989). A Simple Approach to Specifying Concurrent Systems. *Communications of the ACM*, 32:32–45.

Lane, A. (1987). Simulating a Microprocessor in Prolog. *BYTE*, 8:161–168.

Lassez, J.-L., Maher, M. J., and Marriott, K. G. (1986). Unification Revisited. *Computer Science*, 86(12).

Letz, R. (1988). Expressing First-Order Logic with Horn Clause Logic. Technical report, Institut für Informatik, Technische Universität München.

Letz, R., Bayerl, S., Schumann, J., and Bibel, W. (1990). SETHEO – A High-Performance Theorem Prover. *Journal of Automated Reasoning*.

Letz, R. and Schumann, J. M. P. (1988). Global Variables in Logic Programming. Technical report, Institut für Informatik, Technische Universität, München.

Levi, G. (1986). Concurrency Issues in Logic Languages. In *Future Parallel Computers*. Springer.

Levi, G. and Palamidessi, C. (1988). Contributions to the Semantics of Logic Perpetual Processes. *Acta Informatica*, 25:691–711.

Levy, J. (1986a). CFL-A Concurrent Functional Language Embedded in a Concurrent Logic Programming Environment. Technical report, Weizmann Institute of Science, Rehovot, Israel.

Levy, J. (1986b). Shared Memory Execution of Committed-choice Languages. In *Conference On Logic Programming 86*, pages 299–312.

Levy, J. and Friedmann, N. (1986). Concurrent Prolog Implementations - Two New Schemes. Technical report, Weizmann Institute of Science, Rehovot, Israel.

Li, G. and Wah, B. (1985). MANIP-2: A Multicomputer Architecture for Evaluating Logic Programs. *IEEE*, pages 123–130.

Lichtenwalder, K. (1988). Spezifikation einer parallelen Inferenzmaschine in Hinblick auf ein Transputersystem. Master's thesis, Institut für Informatik, Technische Universität München.

Lin, Y.-J. and Kumar, V. (1988). An Execution Model for Exploiting AND-Parallelism in Logic Programs. *New Generation Computing*, 5:393–425.

Liskov, B. (1988). Distributed Programming in Argus. *Communications of the ACM*, 31(3):300–312.

Lloyd, J. (1984). *Foundations of Logic Programming*. Springer, Berlin, Heidelberg.

Lloyd, J. (1989). Meta-programming for knowledge base systems. In Brauer, W. and Freksa, C., editors, *Wissensbasierte Systeme '89*, Berlin. Springer.

Loveland, D. (1968). Mechanical theorem proving by model elimination. *Communications of the ACM*, 15:236–251.

Loveland, D. (1969). A simplified format for the model elimination procedure. *Communications of the ACM*, 16:349–363.

Loveland, D. W. (1979). *Automated Theorem Proving: A Logical Basis*. Elsevier.

Lusk, E., Butler, R., Disz, T., Olson, R., Overbeek, R., Stevens, R., Warren, D. H. D., Calderwood, A., Szeredi, P., Haridi, S., Brand, P., Carlsson, M., Ciepielewski, A., and Haussmann, B. (1988). The AURORA OR-Parallel PROLOG System. In *International Conference on Fifth Generation Computer Systems*, pages 819–830.

Lusk, E. and Overbeek, R., editors (1988). *CADE '88: 9th International Conference on Automated Deduction*, volume 310 of *Lecture Notes in Computer Science*, Berlin. Springer.

Lütke-Holz, B. (1989). Simulation eines parallelen Hornklauselinterpreters nach dem Prinzip der Cooperative Competition. Master's thesis, Institut für Informatik, Technische Universität München.

Maes, P. (1988). Computational reflection. Technical report, AI-Lab, Vrije Universiteit Brussel.

Maes, P. (1989). How to do the right thing. *Connection Science*, 1(3):291–323.

Maes, P. and Nardi, D., editors (1988). *Meta-Level Architectures and Reflection*. North Holland, Amsterdam.

Manna, Z. and Waldinger, R. (1985). *The Logical Basis for Computer Programming Vol. 1: Deductive Reasoning*. Addison-Wesley, Reading, MA.

Männer, R. (1987). *Entwurf und Realisierung eines Multiprozessors*. Informatik-Fachberichte 138. Springer, Berlin.

Mannila, H. and Ukkonen, E. (1986). On the Complexity of Unification Sequences. In *Third International Conference On Logic Programming '86*, pages 122–133.

Martelli, A. and Montanari, U. (1982). An Efficient Unification Algorithm. *ACM Transactions on Programming Languages and Systems*, 4(2):258–282.

Matsuda, H. and Kokata, M. e. a. (1985). Parallel Prolog Machine PARK: Its Hardware Structure and Prolog System. In *Conference on Logic Programming '85*, pages 148–158.

Matsumoto, H. (1985). A Static Analysis of Prolog Programs. *SIGPLAN Notices*, 20(10):48–59.

Maurer, H. (1977). *Theoretische Grundlagen der Programmiersprachen: Theorie der Syntax*. Wissenschaftsverlag / Bibliographisches Institut, Zürich.

Mayr, K. and Reich, M. (1989). Hochparallele Algorithmen zum Erfüllbarkeitsproblem in der Aussagenlogik. Fortgeschrittenenpraktikum, Institut für Informatik, Technische Universität München.

McCarthy, J. (1988). Mathematical logic in artificial intelligence. *Daedalus*, 117 (Winter 1988):297–311.

McClelland, J. L. and Rumelhart, D. E. (1988). *Explorations in Parallel Distributed Processing: A Handbook of Models, Programs, and Exercises*. MIT Press, Cambridge.

McClelland, J. L., Rumelhart, D. E., and the PDP Research Group (1986). *Parallel Distributed Processing: Explorations in the microstructure of cognition*, volume 2 - Psychological and biological models. MIT Press, Cambridge.

McCorduck, P. (1983). Introduction to the Fifth Generation. *Communications of the ACM*, 6(9):629–645.

McCulloch, W. S. and Pitts, W. (1943). A Logical Calculus of the Ideas Immanent in Nervous Activity. *Bulletin of Mathematical Biophysics*, 5:115–133.

Meseguer, J. (1990a). Conditional rewriting logic: Deduction, models and concurrency. Technical Report SRI-CSL-90-14, SRI International, Menlo Park, CA 94025.

Meseguer, J. (1990b). A logical theory of concurrent objects. Technical Report SRI-CSL-90-07, SRI International, Menlo Park, CA 94025.

Miller, D. (1983). Proofs in Higher-Order Logic. Technical report, University of Pennsylvania, Department of Computer and Information Science, Philadelphia.

Mills, J. W. (1990). Connectionist logic programming. Technical Report TR 315, Computer Science Department, Indiana University, Bloomington IN 47405.

Mills, W. (1989). A High-Performance Low Risc Machine for Logic Programming. *Journal of Logic Programming*, pages 179–212.

Milner, R. (1980). *A Calculus of Communicating Systems*, volume 92 of *Lecture Notes in Computer Science*. Springer, Berlin.

Minsky, M. (1981). A framework for representing knowledge. In Haugeland, J., editor, *Mind Design: Philosophy, Psychology, Artificial Intelligence*, pages 95–128. MIT Press, Cambride, MA.

Minsky, M. (1986). *The Society of Mind*. Simon & Schuster, New York.

Minsky, M. (1990). Logical vs. analogical or symbolic vs. connectionist or neat vs. scruffy. In *Frontiers of Artificial Intelligence*, chapter 9, pages 218–243. MIT Press.

Minsky, M. and Papert, S. (1988). *Perceptrons*. MIT Press, Massachusetts. (expanded edition).

Moniz Pereira, L., Monteiro, L., and Cunha, Jose C.and Aparicio, J. (1988). Concurrency and Communication in Delta Prolog. In *IEEE International Specialists Seminar on The Design and Applications of Parallel Digital Processors*, pages 94–104, Lisbon.

Munch, K. H. (1988). A new reduction rule for the connection graph proof procedure. *Journal of Automated Reasoning*, (4):425–444. Automated theorem proving, connection graphs, resolution, order-sorted logic.

Munsch, F. (1989). Ausnutzung von Parallelität bei Theorembeweisern durch Kooperation. Master's thesis, Institut für Informatik, Technische Universität München.

Nagao, M. (1990). *Knowledge and Inference*. Academic Press, San Diego, CA.

Neumayer, G. (1988). Konnektionismus und Aussagenlogik. Master's thesis, Institut für Informatik, Technische Universität München.

Niemilae, I. and Tuominen, H. (1987). Helsinki Logic Machine: A System for Logical Expertise. Technical report, Helsinki University of Technology, Digital Systems Laboratory, Helsinki.

Noske, W. (1989). Entwicklung und Realisierung eines parallelen Debugging-Monitors mit Instant Replay für ein Transputer-Netzwerk. Master's thesis, Institut für Informatik, Technische Universität München.

Ohki, M., Takeuchi, A., and Furukawa, K. (1987). An Object-oriented Language Based on the Parallel Logic Programming Language KL1. In *Conference on Logic Programming '87*, pages 894–909. MIT Press.

Pagels, H. (1988). *The Dreams of Reason*. Simon & Schuster, New York.

Palm, G. (1980). On Associative Memory. *Biological Cybernetics*, 36:19–31.

Palm, G. (1988). Assoziatives Gedächtnis und Gehirntheorie. *Spektrum der Wissenschaft 6/88*, pages 54–64.

Pancake, C. M. and Bergmark, D. (1990). Do parallel languages respond to the needs of scientific programmers? *IEEE Computer*, pages 13–23.

Paola, F. D. (1988). Human-oriented and machine-oriented reasoning: Remarks on some problems in the history of automated theorem proving. *AI & Society*, 2:121–131.

Pappert, S. (1980). *Mindstorms: Children, computers, and powerful ideas*. Basic Books. New York.

Parberry, I. (1987). *Parallel Complexity Theory*. Pitman, London; Wiley, New York.

Park, C.-I., Park, K. H., and Kim, M. (1988). Efficient Backward Execution in and/or Process Model. *Information Processing Letters*, 29:191–198.

Partsch, H. and Moller, B. (1987). Konstruktion korrekter Programme durch Transformation. *Informatik-Spektrum*, 10:309–323.

Paterson, M. and Wegman, M. (1978). Linear Unification. *Journal of Computer and System Sciences*, 16(2):158–167.

Patterson, D. A. (1985). Reduced Instruction Set Computers. *Communications of the ACM*, 28:8–21.

Paul, W. (1978). *Komplexitätstheorie*. B.G. Teubner Verlag.

Pelletier, F. (1986). Seventy-five problems for testing automatic theorem provers. *Journal of Automated Reasoning*, 2:191–216.

Pereira, L. and Nasr, R. (1984). Delta-Prolog: A Distributed Logic Programming Language. In *International Conference On Fifth Generation Computer Systems*, pages 283–291. Institute for New Generation Computer Technology (ICOT).

Perlis, D. (1985). Languages with Self-Reference I: Foundations. *Artificial Intelligence*, 25, Nr. 3:301–322.

Pinegger, T. (1987). *From Equationally Defined Functions to Parallel Processes*. PhD thesis, Fakultät für Mathematik und Informatik, Universität Passau.

Pinkas, G. (1990). Connectionist energy minimization and logic satisfiability. Technical report, Center for Intelligent Computing Systems, Department of Computer Science, Washington University.

Pinkas, G. (1991). Propositional non-monotonic reasoning and inconsistency in symmetric neural networks. In *IJCAI-91*.

Pique, J.-F. (1988). Prolog II: A Step On The Prolog Road. *AICOM*, 1/2:4–16.

Plaisted, D. (1984). The Occur-Check Problem in Prolog. *New Generation Computing*, 2:309–322.

Plaisted, D. A. (1981). Theorem proving with abstraction. *Artificial Intelligence*, (16):47–108.

Plaisted, D. A. (1987). Abstraction using generalization functions. In *CADE '87*, pages 365–376.

Pollack, J. B. (1990). Recursive distributed representations. *Artificial Intelligence*, 46:77–105.

Ponder, C. and Patt, Y. (1984). Alternative proposals for implementing prolog concurrently and implications regarding their respective microarchitectures. In *17th Annual Microprogramming Workshop*.

Pospesel, H. (1976). *Introduction to Logic: Predicate Logic*. Prentice Hall, Eaglewood Cliffs.

Pospesel, H. (1984). *Introduction to Logic: Propositional Logic*. Prentice Hall, Eaglewood Cliffs.

Potter, J. (1985). *The Massively Parallel Processor*. MIT Press.

Prawitz, D. (1970). A proof procedure with matrix reduction. In *Symposium on Automatic Demonstration, Lecture Notes in Mathematics 125*, pages 207–214. Springer, Berlin.

Quinlan, J. (1990). Learning logical definitions from relations. *Machine Learning*, 5:239–266.

Quinton, P. (1987). An introduction to systolic architectures. In (Treleaven and Vanneschi, 1987).

Radig, B., Kodratoff, Y., Überreiter, B., and Wimmer, K.-P., editors (1988). *ECAI 88 – European Conference on Artificial Intelligence*, London. Pitman.

Ramesh, R., Verma, R., Krishnaprasad, T., and Ramakrishnan, I. (1989). Term matching on parallel computers. *Logic Programming*, pages 213–228.

Rapp, W. (1988). PEPSys Sequential Module on the MX-500 Users Manual. Technical Report PEPSys-26, European Computer Research Center (ECRC), München, München.

Ratcliffe, M. and Robert, P. (1986). PEPSys: A Prolog for Parallel Processing. Technical Report CA-17, European Computer Research Center (ECRC), München, München.

Ratcliffe, M. and Syre, J.-C. (1987). Virtual Machines for Parallel Architectures. Technical report, European Computer Research Center (ECRC), München, München.

Refenes, A. (1987). N-expression implementations for integral symbolic and numeric processing. *Future Generation Computer Systems*, 3:161–187.

Richards, B. (1988). When Facts Get Fuzzy. *BYTE*, (4):285–290.

Ringwood, G. (January 1988). Parlog86 and the Dining Logicians. *Communications of the ACM*, 31:10–25.

Robinson, I. (1986). A Prolog Processor Based on a Pattern Matching Memory Device. In *Third International Conference On Logic Programming '86*, pages 172–179.

Robinson, J. (1965). A machine-oriented logic based on the resolution principle. *Journal of the ACM*, 12:23–41.

Robinson, J. (1968). The generalized resolution principle. *Machine Intelligence*, 3:77–94.

Robinson, J. (1979). *Logic: form and function*. Edinburgh University Press, Scotland and Elsevier-North Holland, New York.

Rohmer, J., Gonzalez-Rubio, R., and Bradier, A. (1986). Delta driven computer: A parallel machine for symbolic processing. In (Treleaven and Vanneschi, 1987).

Rumelhart, D. E., McClelland, J. L., and the PDP Research Group (1986). *Parallel Distributed Processing: Explorations in the microstructure of cognition*, volume 1 - Foundations. MIT Press, Cambridge.

Sacerdoti, E. D. (1975). Planning in a hierarchy of abstraction spaces. *IJCAI*, pages 412–422.

Safra, S. (1986). Partial Evaluation of Concurrent Prolog and its Implication. Technical Report CS86-24, Weizmann Institute of Science.

Sand, S. (1986). *Künstliche Intelligenz – Geschichten über Menschen und denkende Maschinen*. Heyne Verlag, München.

Saraswat, V. (1986). Problems with Concurrent Prolog. Technical Report CME-CS-86-100, Carnegie-Mellon University.

Schaller, H. N. (1990). Solving constraint problems using feedback neural networks. In (Eckmiller et al., 1990), pages 265–268.

Schank, R. C. and Childers, P. (1984). *The Cognitive Computer: On Language, Learning and Artificial Intelligence*. Addison-Wesley, Massachusetts.

Schmid, E. (1988). Implementierung eines parallelen Theroembeweisers auf einem Multiprozessor-Simulator. Fortgeschrittenenpraktikum für informatiker, Institut für Informatik, Technische Universität München, München.

Schnoebelen, P. (1986). $\mu$ - FP2. A prototype interpreter for FP2. Technical Report 573, LIFIA-IMAG, Grenoble, France.

Schnoebelen, P. (1987). Rewriting techniques for the temporal analysis of communicating processes. In *PARLE '87*, volume 259 of *Lecture Notes in Computer Science*. Springer. Volume 2.

Schnoebelen, P. (1988). Refined compilation of pattern matching for functional languages. *Science of Computer Programming*, 11(2):133–159.

Schnoebelen, P. and Jorrand, P. (1989). Principles of FP2: term algebras for specification of parallel machines. In De Bakker, J., editor, *Languages for Parallel Architectures: Design, Semantics, Implementation Models*. Wiley.

Schnorr, C. (1974). *Rekursive Funktionen und ihre Komplexität*. Teubner Studienbücher Informatik. B.G. Teubner, Stuttgart.

Schöning, U. (1987). *Logik für Informatiker*. Reihe Informatik, Band 56. Wissenschaftsverlag / Bibliographisches Institut, Zürich.

Schumann, J. (1991). *Efficient Theorem Provers based on an Abstract Machine*. PhD thesis, Institut für Informatik, Technische Universität München.

Schumann, J., Ertel, W., and Suttner, C. (1989). Learning heuristics for a theorem prover using back propagation. In *ÖGAI 89*.

Schumann, J. and Letz, R. (1990). PARTHEO: A High Performance Parallel Theorem Prover. In (Stickel, 1990b), pages 40 – 56.

Schwaab, F. and Tusera, D. (1988). Un Algorithme Distribue pour l'Execution Parallele de Prolog. Technical report, INRIA, Le Chesnay.

Seitz, C. L. (1984). Concurrent VLSI Architectures. *IEEE Transactions on Computers*, c-33(12):1247–1265.

Shafer, G. and Pearl, J., editors (1990). *Readings in Uncertain Reasoning*. Representation and Reasoning. Morgan Kaufmann, San Mateo, CA.

Shapiro, E. (1983). A Systolic Concurrent PROLOG Machine – Lecture notes on the Bagel. Technical Report TR-035, Institute for New Generation Computer Technology (ICOT), Tokyo.

Shapiro, E. (1984). Systolic Programming: A Paradigm of Parallel Processing. *International Conference on Fifth Generation Computer Systems*, pages 458–470.

Shapiro, E. (1986). Concurrent prolog: A progress report. *Computer*, 1986(8):44–58; also in (Bibel and Jorrand, 1986), pages 277–313

Shapiro, E. (1988). *Concurrent Prolog*. MIT Press.

Shapiro, E. (1989a). The family of concurrent logic programming languages. *ACM Computing Surveys*, 21(3):413–510.

Shapiro, E. (1989b). Or-Parallel PROLOG in Flat Concurrent PROLOG. *Logic Programming*, (6):243–267.

Shapiro, E. and Takeuchi, A. (1983). Object-Oriented Programming in Concurrent Prolog. *New Generation Computing*, (1):25–48.

Shastri, L. (1988). A connectionist approach to knowledge representation and limited inference. *Cognitive Science*, 12:331–392.

Shastri, L. and Ajjanagadde, V. (1989). A connectionist system for rule based reasoning with multi-place predicates and variables. Technical report, University of Pennsylvania, Computer and Information Science Department, Philadelphia.

Shastri, L. and Ajjanagadde, V. (1990). From simple associations to systematic reasoning: A connectionist representation of rules, variables and dynamic bindings. Technical Report MS-CIS-90-05, Computer And Information Science Department, University of Pennsylvania, Philadelphia, PA 19104.

Shastri, L. and Feldman, J. (1985). Evidential Reasoning in Semantic Networks: A Formal Theory. In *IJCAI '85*, pages 465–474.

Shastri, L. and Feldman, J. (1986). Neural nets, routines, and semantic networks. In Sharkey, N., editor, *Advances in Cognitive Science 1*, pages 158–203.

Shaw, D. (1981). NON-VON: A Parallel Machine Architecture for Knowledge Based Information Processing. In *IJCAI '81*, pages 961–963, Vancouver.

Shaw, D. (1987). On the range of applicability of an artificial intelligence machine. *Artificial Intelligence*, 32:252–172.

Shen, K. and Warren, D. (1987). A simulation study of the Argonne model for OR-parallel execution of Prolog. In *Symposium on Logic Programming '87*, pages 54–86.

Shrobe, H., Aspinall, J., and Mayle, N. (1988). Towards A Virtual Parallel Inference Engine. In *AAAI '88*, pages 654–659.

Siekmann, J. and Wrightson, B., editors (1983). *Automation of Reasoning*, volume 1, 2. Springer, Berlin.

Silver, B. (1986). *Meta-level Inference*. North Holland.

Singhal, A. (1990). *Exploiting Fine Grain Parallelism in Prolog*. PhD thesis, University of California, Berkeley. TR CSD 90/588.

Sinowjew, A. and Wessel, H. (1975). *Logische Sprachregeln: Eine Einführung in die Logik*. Wilhelm Fink Verlag, München-Salzburg.

Smith, B. (1984). Logic Programming on an FFP Machine. In *International Symposium on Logic Programming*, Atlantic City, New Jersey.

Smullyan, R. (1968). *First-Order Logic*. Springer, Berlin.

Sohma, Y., Satoh, K., Kumon, K., Masuzawa, H., and Itashiki, A. (1985). A new parallel inference mechanism based on sequential processing. In *IFIP TC-10 Working Conference on Fifth Generation Computer Architecture*, UMIST, Manchester.

Srini, V. P., Tam, J., Nguyen, T., Chen, C., Wei, A., Testa, J., Patt, Y. N., and Despain, A. M. (1987). VLSI Implementation of A Prolog Processor. In *Stanford VLSI Conference*.

Stanfill, C. and Waltz, D. (1986). Toward Memory-Based Reasoning. *Communications of the ACM*, 29:1213–1228.

Stender, J. (1987). Parallele Prolog-Implementierung auf Transputern. *Hard and Soft*, Juli/August 87:17–23.

Sterling, L. and Shapiro, E. (1986). *The Art of Prolog. Advanced Programming Techniques*. MIT Press, Cambridge, MA.

Stern, A. (1988). *Matrix Logic*. North Holland, Amsterdam.

Sternberg, R. J. (1982). Reasoning, problem solving, and intelligence. In Sternberg, R. J., editor, *Handbook of Human Intelligence*, chapter 5, pages 225–307. Cambridge University Press, Cambridge.

Stickel, M. (1984). A Prolog Technology Theorem Prover. *New Generation Computing*, 2:371–383.

Stickel, M. (1986). An Introduction to Automated Deduction. In Bibel, W. and Jorrand, P., editors, *Fundamentals of Artificial Intelligence*, volume 232 of *Lecture Notes in Computer Science*, pages 75–132. Springer.

Stickel, M. (1988). A prolog technology theorem prover: Implementation by an extended prolog compiler. *Journal of Automated Reasoning*, (4):353–380. Automated theorem proving, model elimination procedure, Prolog.

Stickel, M., editor (1990). *CADE '90: 10th International Conference on Automated Deduction*, volume 449 of *Lecture Notes in Artificial Intelligence*, Kaiserslautern. Springer.

Stolcke, A. (1988). Generierung natürlichsprachlicher Sätze in unifikationsbasierten Grammatiken – Ein Konnektionistischer Ansatz. Master's thesis, Institut für Informatik, Technische Universität München.

Stolcke, A. (1989). Unification as constraint satisfaction in structured connectionist networks. *Neural Computation*, (1):559–567.

Stolfo, S. (1983). The DADO Parallel Computer. Technical report, Department of Computer Science, Columbia University, New York.

Stolfo, S. (1987). Initial Performance of the DADO-2 Prototype. *Computer*, 20:75–85.

Stolfo, S., Miranker, D., and Shaw, D. (1983). Architecture and applications of DADO: A large-scale parallel computer for artificial intelligence. In *IJCAI '83*, pages 850–854, Karlsruhe, BRD.

Suttner, C. B. (1989). Learning Heuristics for Automated Theorem Proving. Master's thesis, Institut für Informatik, Technische Universität München.

Swain, M. and Cooper, P. (1988). Parallel Hardware for Constraint Satisfaction. In *AAAI '88*, pages 682–686.

Syre, J.-C. (1985). A Review of Computer Architectures for Functional and Logic Programming Systems. Technical report, European Computer Research Center (ECRC), München, Munich, Germany.

Syre, J.-C. and Westphal, H. (1985). A Review of Parallel Models for Logic Programming Languages. Technical report, European Computer Research Center (ECRC), München, München.

Szpakowicz, S. (1987). Logic Grammars. *BYTE*, (8):185–193.

Takeuchi, A. and Furukawa, K. (1985). Interprocess Communication in Concurrent Prolog. Technical report, Institute for New Generation Computer Technology (ICOT), Tokyo.

Takeuchi, A. and Furukawa, K. (1986). Parallel Logic Programming Languages. In *Third International Conference On Logic Programming '86*, pages 242–254.

Takeuchi, A., Takahashi, K., and Shimizu, H. (1987). A Description Language with AND/OR Parallelism for Concurrent Systems and Its Stream-Based realization. Technical report, Institute for New Generation Computer Technology, Tokyo.

Taki, K. (1985). ICOT Progress Report. *IFIP TC-10 Conference on Fifth Generation Computer Architecture*. Research plans for parallel inference machine 1985-1988.

Taki, K., Yokota, M., Yamamoto, A., Nishikawa, H., Uchida, S., Nakashima, H., and Mitsuishi, A. (1984). Hardware Design and Implementation of the Personal Sequential Inference Machine (PSI). In *International Conference on Fifth Generation Computer Systems '84*. Institute for New Generation Computer Technology (ICOT).

Tamaki, H. (1985). A Distributed Unification Scheme for Systolic Logic Programs. pages 552–559. IEEE.

Tamura, N. and Kanada, Y. (1984). Implementing Parallel Prolog on a Multiprocessor Machine. In *International Symposium On Logic Programming '84*, Atlantic City, NJ.

Tanaka, J. (1988). Meta-interpreters and Reflective Operations in GHC. In *Future Generation Computer Systems*, pages 775–783, Tokyo. Institute for New Generation Computer Technology (ICOT).

Taylor, S. (1989). *Parallel Logic Programming Techniques*. Prentice Hall.

Taylor, S., Maio, C., Stolfo, S., and Shaw, D. (1983). Prolog on the DADO Machine: A Parallel System for High-speed Logic Programming. Technical report, Department of Computer Science, Columbia University, New York.

Terrano, A. E., Dunn, S. M., and Peters, J. E. (1989). Using an architectural knowledge base to generate code for parallel computers. *Communications of the ACM*, 32(9):1065–1072.

Thinking Machines (1987). Connection Machine Model CM-2 Technical Summary. Technical Report HA87-4, Thinking Machines Corporation.

Tick, E. and Warren, D. (1984). Towards a Pipelined Prolog Processor. *New Generation Computing*, 2:323–345.

Tokoro, M. and Ishikawa, Y. (1986a). Concurrent Programming in Orient 84K: An Object-Oriented Knowledge representation Language. *SIGPLAN Notice*, 21:2–10.

Tokoro, M. and Ishikawa, Y. (1986b). Orient 84/K: A Language within Multiple Paradigms in the Object Framework. In *19th Annual Hawaii International Conference On System Science*, pages 198–207.

Touretzky, D. and Hinton, G. (1985). Symbols Among the Neurons: Details of a Connectionist Inference Architecture. In *IJCAI '85*, Pittsburgh.

Touretzky, D. S. and Hinton, G. E. (1988). A distributed connectionist production system. *Cognitive Science*, 12:423–466.

Trapp, N. (1987). Strategies and Heuristics for Theorem Provers. Master's thesis, Institut für Informatik, Technische Universität München.

Treleaven, P. C., editor (1989). *Parallel Computers: Object-Oriented, Functional and Logic*. Wiley, Chichester.

Treleaven, P. C. and Refenes, A. N. (1985). Fifth Generation and VLSI Architectures. *Future Generation Computer Systems*, 1(6):387–396.

Treleaven, P. C., Refenes, A. N., Lees, K. J., and Mccabe, S. C. (1986). Computer Architectures for Artificial Intelligence. Technical report, University College, London; Ferranti Computer Systems, Bracknell; THORN-EMI Central Research Labs., Hayes.

Treleaven, P. C. and Vanneschi, M., editors (1987). *Future Parallel Computers*, volume 272 of *Lecture Notes in Computer Science*, Berlin. Springer.

Tucker, L. W. and Robertson, G. G. (1988). Architecture and Applications of the Connection Machine. *Computer*, (8):26–38.

Uchida, S. (1987). Parallel Inference Machines at ICOT. *Future Generation Computer Systems*, (3):245–252.

Uchida, S., K, T., Goto, A., Nakajima, K., Nakashima, H., Yokota, M., Nishikawa, H., Yamamoto, A., and Mitsui, M. (1986). Logic Computers and Japan's FGCS Project. In (Treleaven and Vanneschi, 1987).

Uchida, S., Taki, K., Nakajima, K., Goto, A., and Chikayama, T. (1988). Research and Development of the Parallel Inference Systemin the Intermediate Stage of the FGCS Project. In *International Conference on Fifth Generation Computer Systems*, pages 17–36, Tokyo. Institute for New Generation Computer Technology (ICOT).

Ueda, K. (1985). Guarded Horn Clauses. Technical Report TR-103, Institute for New Generation Computer Technology (ICOT).

Ueda, K. (1986). Guarded Horn Clauses: A Parallel Logic Programming Language with the Concept of a Guard. Technical Report TR-208, Institute for New Generation Computer Technology (ICOT), Tokyo, Japan.

Ueda, K. (1989). Parallelism in logic programming. Technical Report TR-495, Institute for New Generation Computer Technology (ICOT), Tokyo, Japan.

Ultsch, A., Hannuschka, R., Hartmann, U., and Weber, V. (1990). Learning of control knowledge for symbolic proofs with backpropagation networks. In (Eckmiller et al., 1990), pages 499–502.

Ungerer, T. and Zehendner, E. (1988a). A Parallel Computer Architecture Directed Towards Modular Concurrent Programming. Technical Report 177, Universität Augsburg, Institut für Mathematik.

Ungerer, T. and Zehendner, E. (1988b). Language Abstractions for Concurrency Control. Technical report, Universität Augsburg, Institut für Mathematik.

van Emden, M. and Kowalski, R. (1976). The Semantics of Predicate Logic as a Programming Language. *Journal of the Association for Computing Machinery*, 23, No. 4:733–742.

van Emden, M. and Lloyd, J. (1984). A Logical Reconstruction of Prolog II. *J. Logic Programming*, 2:143–149.

Van Roy, P. (1991). The Aquarius Project. Technical report, Computer Science Department, University of California, Berkeley.

Vlahavas, J. and Halatsis, C. (1989). L-Machine: A Low-Cost Personal Sequential Inference Machine. *Journal of Systems and Software*, (9):209–223.

Voda, P. and Yu, B. (1984). RF-Maple: A Logic Programming Language with Functions, Types, and Concurrency. In *International Conference On Fifth Generation Computer Systems*, pages 341–347.

von Neumann, J. (1958). The Computer and the Brain. In (Anderson and Rosenfeld, 1988), pages 81–89. from: The Computer and the Brain; New Haven: Yale University Press 1958, pp.66-82.

Wah, B. (1987). New Computers for Artificial Intelligence Processing. *Computer*, 20:10–19.

Wallace, M. (1989). A computable semantics for general logic programs. *Journal of Logic Programming*, (6):269–297.

Waltz, D. (1987). Applications of the Connection Machine. *Computer*, 20:85–100.

Waltz, D. and Feldman, J., editors (1988). *Connectionist Models and Their Implications*. Ablex Publishing, Norwood, NJ.

Waltz, D. and Stanfill, C. (1988). Artificial Intelligence Related Research on the Connection Machine. In *International Conference on Fifth Generation Computer Systems*, pages 1010–1024, Tokyo. Institute for New Generation Computer Technology (ICOT).

Wang, J. (1989). Towards a New Computational Model for Logic Languages. Technical report, Department of Computer Science, University of Essex, Colchester.

Warren, D. (1987). The SRI model for OR-parallel execution of Prolog. In *Symposium on Logic Programming '87*, pages 92–102.

Warren, D. H. (1983). An Abstract Prolog Instruction Set. Technical Report 309, SRI International, Artificial Intelligence Center, Menlo Park, California.

Watzlawik, G., Benker, H., and Noyé, J. (1987). ICM 4. Technical Report CA-25, European Computer Research Center (ECRC), München, München.

Weinbaum, D. and Shapiro, E. (1986). Hardware Description and Simulation Using Concurrent Prolog. Technical Report CS86-25, Weizmann Institute of Science, Rehovot, Israel.

Westphal, H. (1986). Eine Beurteilung paralleler Modelle für Prolog. In *GI-Jahrestagung '86*, pages 227–240, Berlin. Springer.

Westphal, H., Robert, P., Chassin, J., and Syre, J. (1987). The PEPSys model: Combining Backtracking, AND- and OR-parallelism. In *Symposium on Logic Programming '87*, pages 436–448.

Weyrauch, R. (1980). Prolegomena to a Theory of Mechanized Formal Reasoning. *Artificial Intelligence*, 13:133–170.

White, H. (1989). Learning in artificial neural networks: A statistical perspective. *Neural Computation*, 1:425–464.

Whitehead, A. and Russell, B. (1910). Principia Mathematica. Cambridge.

Widrow, B. and Hoff, M. (1960). Adaptive switching circuits. In (Anderson and Rosenfeld, 1988), pages 123–134. from: 1960 IRE WESCON Convention Record, New York: IRE, pp.96-104.

Wiener, N. (1948). *Cybernetics*. Wiley, New York.

Wittgenstein, L. (1921). Tractatus Logico-Philosophicus. London.

Wolfe, M. and Banerjee, U. (1987). Data Dependence and its Application to Parallel Processing. *Parallel Programming*, 16:137–178.

Wolfram, D., Maher, M., and Lassez, J.-L. (1984). A unified treatment of resolution strategies for logic programs. In *Second International Logic Programming Conference*, Uppsala University, Schweden.

Wos, L., Overbeek, R., Lusk, E., and Boyle, J. (1984). *Automated Reasoning: Introduction and Applications*. Prentice-Hall, Eaglewood Cliffs.

Yasuura, H. (1984). On Parallel Computational Complexity of Unification. In *International Conference on Fifth Generation Computer Systems*, pages 235–243. Institute for New Generation Computer Technology (ICOT).

Zadeh, L. (1983). The Role of Fuzzy Logic in the Management of Uncertainty in Expert Systems. *Fuzzy Sets and Systems*, (11):199–227.

# Index

# Automated Theorem Proving

by Wolfgang Bibel

*2nd revised edition 1987. xiv, 289 pp. (Artificial Intelligence, ed. by Wolfgang Bibel) Softcover.*
*ISBN 3-528-18520-1*

In modern computing technology there is a clear trend towards more flexible and more intelligent systems. A key feature for achieving such flexibility is the capability for sophisticated and efficient deductive reasoning. This book provides a comprehensive treatment of the current state of the art in inferential technology based on classical logic. Therefore it provides valuable information for many areas in computing science. For instance, structuring and processing knowledge (e.g. for expert system) as well as programming and problem solving are among the applications treated in the text.

The revised second edition brings the book back to the current state of the art.

Verlag Vieweg · Postfach 58 29 · D-6200 Wiesbaden 1